EXCEL MANUAL
for Moore, McCabe, Duckworth, and Sclove's

# The Practice of Business Statistics

Fred M. Hoppe
*McMaster University*

W. H. Freeman and Company
New York

Printed in the United States of America

ISBN: 0-7167-6640-X

First printing 2003

W. H. Freeman and Company
41 Madison Avenue
New York, NY 10010
Houndmills, Basingstoke
RG21 6XS England

# Contents

# Preface

This book is a supplement to the first edition of *The Practice of Business Statistics (PBS)* by Moore, McCabe, Duckworth, and Sclove. Its purpose is to present Excel as an approach to performing common statistical procedures.

Excel is a spreadsheet program, a tool for organizing data contained in columns and rows. Operations on data mimic those described by mathematical functions, and the formulas required for data analysis can therefore be expressed as spreadsheet operations. Although originally developed as a business application to display numbers in a tabular format and to automatically recalculate values in response to changes in numbers in the table, current spreadsheets enjoy built-in functions and display capabilities that can be used for statistical analysis.

Excel is also part of an integrated word processing/graphics/database package, and therefore data can be input easily from other applications and then exported into report form. Excel is used in many business courses and therefore facility with Excel that is developed becomes a useful skill that can be used by students after graduation. For instance, I regularly download stock prices into Excel from the Internet to plot the course of investments, and my students enjoy this particular application as an incentive for learning statistics. Even non-business students who learn statistics using Excel will find it extremely valuable in future courses (a point brought home by one of my third-year students who informed me that she used Excel in her physics course for plotting data because she "didn't feel like doing it by hand").

For the instructor, Excel is an ideal alternative to specialized statistical software for teaching purposes. Most students have had prior exposure to Excel and do not view it as a new piece of software to learn. This is a tremendous benefit for the instructor because students arrive in class already knowledgeable in basic data management, and this allows the instructor to concentrate on statistics from the first lecture.

I believe that Excel can satisfy all of a student's needs in a course based on a book such as *PBS*. In fact, every technique discussed in *PBS* can be developed within Excel. Given that Excel is produced by Microsoft, one can predict without fear, a long, useful, and upgradeable life for Excel. All these considerations lead me to believe that Excel will grow in popularity in the teaching of statistics.

This manual is designed along the lines of my earlier ones for *Introduction to the*

*Practice of Statistics* by Moore and McCabe, and *The Basic Practice of Statistics* by Moore, but differs in one important aspect: **Macros**. Included on the CD that accompanies *PBS* are over a dozen macros that have been developed by the author to expand and extend Excel's capabilities. Instructions for using these macros are in the manual. As well, additional macros are planned and updates will be available online to adopters.

Some key features of this book are:

- Introductory chapter on Excel for those with no prior knowledge.
- Presentation follows *PBS* completely with fully worked and cross-referenced examples and exercises from *PBS*.
- Detailed exposition of the ChartWizard for graphical displays.
- Written for all versions of Excel with parallel instructions where there are major differences in the interface.
- Completely integrated for use in Macintosh or Windows environments with equivalent keyboard and mouse commands presented as needed for each operating system.
- Extensive use of Named Ranges to make formulas more transparent.
- More than 200 figures.
- Figures from different versions of Excel showing differences and similarities in the user interface.
- Detailed use of simulation to explain randomness.
- At least one completely worked example with step-by-step Excel instructions for each technique discussed in *PBS*.
- Templates provided for nearly every technique.
- Reference to Microsoft URLs for updates.

This manual may be adopted by users of Excel 97, 2000, and 2002 (Windows) and Excel 98, 2001, and Excel X (Macintosh). Excel 2000/2001 did not make substantial changes from Excel 97/98. There were improvements in the user interfaces and in some features such as the PivotTable, but the main changes dealt with VBA programming and integration with the Internet, for instance in allowing workbooks to be saved as web pages (html files). Users of the recent version of Excel 2002, part of the Office XP suite, should also be fine because again only minor changes were implemented in Excel 2002. Also, there are separate instructions for those who still use the older Excel 5/95.

The people at W.H. Freeman and Co. who were involved in this project were: Patrick Farace, Mark Santee, Danielle Swearengin, Brian Donnellan, and Tim Robinson. My thanks to all.

I will be putting up a Web site at http://www.mathematics.net/ where updates, macros, and other useful information will appear.

Finally, thanks to my loving wife, Marla, for patiently putting up with my involvement in this book and for proofreading the entire manuscript and finding many typos I missed. All remaining errors are my own responsibility. My wonderful and dear children, Daniel and Tamara, helped as much as they could and enjoyed being part of the production process.

FRED M. HOPPE
DUNDAS, ONTARIO
MARCH 15, 2003
E-mail: *hoppe@mcmaster.ca*
http://www.mathematics.net/

# Introduction

This book is a supplement to *The Practice of Business Statistics*, by David S. Moore, George P. McCabe, William M. Duckworth, and Stanley L. Sclove, referred to as *PBS*. Its purpose is to show how to use Excel in performing the common statistical procedures for students of business statistics in *PBS*.

## I.1   What Is Excel?

Microsoft Excel is a spreadsheet application whose capabilities include graphics and database applications. A spreadsheet is a tool for organizing data. Originally developed as a business application for displaying numbers in a table, numbers that were linked by formulas and updated whenever any part of the data in the spreadsheet changed, Excel now has built-in functions, tools, and graphical features that allow it to be used for sophisticated statistical analyses.

### Windows or Macintosh?

It doesn't matter which you use. This book is designed equally for Macintosh or Windows operating systems. The Macintosh and Windows versions of Excel function essentially the same way, with a few slight differences in the file, print, and command shortcuts. These are due mainly to the absence of a right mouse button for the Macintosh. However, the right button action can be duplicated with a keystroke, and I have described both actions where they differ. Both Macintosh and Windows users will find this book useful.

Nearly all figures shown in this book have been generated using Excel 2001 on a Macintosh G4 running Mac OS 10.1.4. Corresponding figures from the Windows version are only cosmetically different, for instance in Fig 1.1 in Chapter 1 showing the opening screen of the ChartWizard. Students should feel familiar with the look of the Excel interface no matter what the platform.

### Which Version of Excel Should I Use?

Naming conventions are slightly confusing because of a plethora of patches, bug fixes, interim releases, and so on. In the Windows environment the main versions used are Excel 5.0, 5.0c, Excel 7.0, 7.0a (for Windows 95), Excel 8.0 (also called

Excel 97), Excel 2000, and Excel 2002. For Macintosh, they are Excel 5.0 and 5.0a, Excel 98, Excel 2001, and Excel X for Mac X.

The major change in the development occurred with Excel 5.0. The statistical tools in Excel 5.0 and Excel 7.0 for Windows and Excel 5.0 for Macintosh function in virtually the same way. These are referred to collectively in this book as Excel 5/95. Likewise, the Windows versions Excel 97, Excel 2000 and the Macintosh versions Excel 98, Excel 2001 function similarly and are referred to as Excel 97/98/2000/2001.

Excel 97/98 removed some bugs in the Data Analysis ToolPak (but introduced others), improved the interface to the ChartWizard, and replaced the Function Wizard with the Formula Palette. Help was greatly expanded with the introduction of the animated Office Assistant. Minor cosmetic changes were made in the placement of components within some dialog boxes. Although changes were made in the VBA interface, few substantive changes occurred that might cause differences in execution or statistical capabilities between Excel 5/95 and Excel 97/98. Both Excel 2000 and Excel 2002 introduced only incremental improvements, mostly in the user interface rather than programming.

This book is based on Excel 2000/2001, but all the instructions, however, have been tested on Excel 97/98 as well as Excel 2002. Moreover, given the existing base of Excel 5/95 users I have tried to make this book equally accessible to them without having two separate editions. Surprisingly, this has not been too difficult. The main differences needing instructional care are the description of the ChartWizard (reduced to four steps in Excel 97/98 from five steps in Excel 5/95) and the implementation of formulas with the Formula Palette (Excel 97/98) instead of the Function Wizard (Excel 5/95). In addition, there are related differences in some pull-down options from the Menu Bar. On the other hand, the Data Analysis ToolPak is virtually the same in each version. Nearly all differences are in Chapters 1 and 2. While earlier editions of this manual included a detailed Appendix covering these two chapters for Excel 5/95, it has been removed since a user of Excel 5/95 can easily make modifications based on that older interface. In the remaining chapters, whenever there is a technique whose implementation varies between the two versions, I have given separate steps. Usually only a few lines of text (four or fewer) suffice, underscoring the similarity in the versions in using Excel for statistics. It is hoped that this dual approach will make this manual truly equally useful for all versions from Excel 5/95 to Excel 2000/2001.

## Do I Need Prior Familiarity with Excel?

The short answer is no. This book is completely self-contained. This introductory chapter contains a summary and description of Excel that should provide enough detail to enable a student to get started quickly in using Excel for statistical calculations. The subsequent chapters give step-by-step details on producing and embellishing graphs, using functions, and invoking the **Analysis Toolbox**. For users without prior exposure to Excel, this book may serve as a gentle exposure to spreadsheets and a starting point for further exploration of their features.

## I.2 The Excel Workbook

When you first open Excel, a new file is displayed on your screen. Fig I.1 shows an Excel 2001 Macintosh opening workbook.

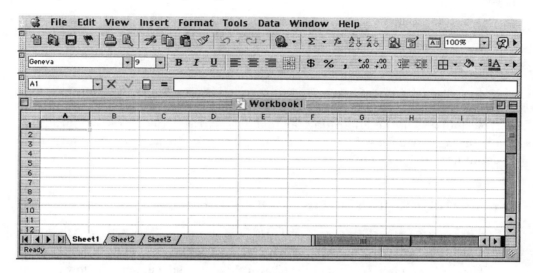

Figure I.1: Excel 2001 Macintosh Workbook

A workbook consists of various sheets in which information is displayed, usually related information such as data, charts, or macros. Sheets may be named and their names will appear as tabs at the bottom of the workbook. Sheets may be selected by clicking on their tabs and may be moved within or between workbooks. To keep the presentation simple in this book we have chosen to use one sheet per workbook in each of our examples.

A sheet is an array of cells organized in rows and columns. The rows are numbered from 1 to 65,536 (up from 16,384 in Excel 5/95) while the columns are described alphabetically, as follows,

$$A, B, C, \ldots, X, Y, Z, AA, AB, AC, \ldots, IV$$

for a total of 256 columns.

Each cell is identified by the column and row that intersect at its location. For instance, the selected cell in Fig I.1 has address (or cell reference) A1. When referring to cells in other sheets, we need to also provide the sheet name. Thus Sheet2!D9 refers to cell D9 on Sheet 2.

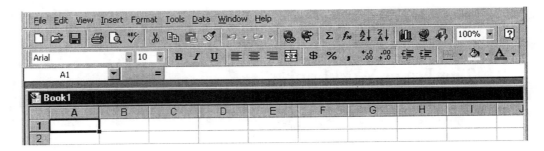

Figure I.2: Excel 97 Windows

Information is entered into each cell by selecting an address (use your mouse or arrow keys to navigate among the cells) and entering information, either directly in the cell or else in the **Formula Bar** text entry area. Three types of information can be entered: labels, values, and formulas. We will discuss entering information in detail later in this chapter.

## I.3   Components of the Workbook

Look at Fig I.1. Then compare with Fig I.2 showing an Excel 97 (Windows) workbook and Fig I.3, an Excel 5 Macintosh workbook.

There are two main components in a workbook: the document window and the application window. Information is entered in the document window identified by the row and column labels. Above the document window are all the applications, functions, tools, and formatting features that Excel provides. There are so many commands in a workbook that an efficient system is needed to access them. This is achieved either by pull-down menus invoked from the Menu Bar or from equivalent icons on the toolbars.

The application window is thus the control center from which the user gives instruction to Excel to operate on the data in the document window below it. We

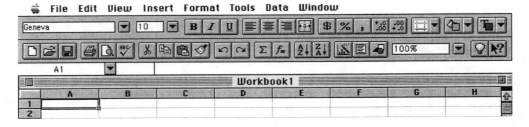

Figure I.3: Excel 5 Macintosh

File  Edit  View  Insert  Format  Tools  Data  Window  Help

Figure I.4: Menu Bar

Table I.1: Menu Bar Pull-Down Options

| Menu Bar | Pull-Down Options |
|----------|-------------------|
| **File** | Open, close, save, print, exit |
| **Edit** | Copy, cut, paste, delete, find, etc. (basic editing) |
| **View** | Controls which components of workbook are displayed on screen, size, etc. |
| **Insert** | Insert rows, columns, sheets, charts, text, etc. into workbook |
| **Format** | Format cells, rows, columns |
| **Tools** | Access spelling macros, data analysis toolpak (will be used throughout to access the statistical features of Excel) |
| **Data** | Database functions such as sorting, filtering |
| **Window** | Organize and display open workbooks |
| **Help** | Online help (also available on the **Standard Toolbar**) |

examine the main components of the application window in detail.

## Menu Bar

The **Menu Bar** (Fig I.4) appears at the top of the screen. It provides access to all Excel commands: **File, Edit, View, Insert, Format, Tools, Data, Window, Help**. Each word in the Menu Bar opens a pull-down menu of options familiar to users of any window-based application (there are also keyboard equivalents). As the name implies, this is the main component of the control center and will be elaborated on in various examples throughout this book. Table I.1 summarizes some of the options available.

## Toolbars

When Excel is opened, two strips of icons appear below the Menu Bar: the **Standard Toolbar** and the **Formatting Toolbar**. Other toolbars are available by choosing **View – Toolbars** from the Menu Bar and making a selection from the choices available. Existing toolbars can be customized by adding or re-

Figure I.5: Excel 2001 Standard Toolbar

moving buttons, and new ones can be created. For the purpose of this book you will not need to make such customizations.

**Standard Toolbar**

The (default) **Standard Toolbar** (Fig I.5) provides buttons to ease your access to basic workbook tasks. Included are buttons for the following tasks:

- Start a new workbook
- Open existing workbook
- Save open workbook
- Print, print preview
- Check spelling
- Cut selection and store in clipboard for posting elsewhere
- Copy selected cells to clipboard, paste data from clipboard
- Copy format
- Undo last action, redo last action
- Insert hyperlink
- World Wide Web interface
- Autosum function (may also be entered directly into cell or **Formula Bar**)
- **Paste Function** (step-by-step dialog boxes to enter a function connected to the **Formula Palette**)
- Sort descending, sort ascending order
- **ChartWizard** (covered in detail in Chapter 1)
- **PivotTable Wizard**
- Drawing toolbar
- Zoom factor for display
- **Office Assistant**

As you pass over a button with your mouse pointer a small text label appears next to the button.

**Formatting Toolbar**

Fig I.6 shows the **Formatting Toolbar**, by which you can change the appearance of text and data. Features offered:

- Display and select font of selected cell

- Apply bold, italic, underline formatting
- Left, center, or right justify data
- Merge and center
- Apply currency style, percent style, etc.
- Increase or decrease decimal places
- Indent
- Add borders to selected sides of cell
- Change background color of cell, change color of text in cells

Figure I.6: Formatting Toolbar

## Formula Bar

The **Formula Bar** is located just above the document window (Fig I.7). There are six areas in the Formula Bar (from left to right):

Figure I.7: Formula Bar

- **Name box.** Displays reference to active cell or function.
- **Defined name pull-down.** Lists defined names in workbook.
- **Cancel box.** Click on the red X to delete the contents of the active cell.
- **Enter box.** Click the green check mark to accept the formula bar entry.
- **Formula Palette (Excel 97/98 and Excel 2000/2001).** Constructs a function using dialog boxes to access Excel's built-in functions, or the function can be entered directly if you know the syntax. (Excel 5/95 has $f_x$ in place of the equal (=) sign to activate the **Function Wizard**).
- **Text/Formula Entry area.** Enter and display the contents of the active cell.

The **Cancel** box and **Enter** box buttons appear only when a cell is being edited. Once the data have been entered, they disappear.

## Title Bar

This is the name of your workbook. On a Mac the default is **Workbook1**, which appears just above the document window. With Windows the default name is **Book1**.

### Document (Sheet) Window

A sheet in a workbook contains 256 columns by 65,536 rows. Use the mouse pointer or arrow keys to move from cell to cell. The pointer may change appearance depending on what actions are permitted. It might be an arrow, a blinking vertical cursor (I-beam), or an outline plus sign, for instance.

### Sheet Tabs

A single workbook can have many sheets, the limit determined by the capacity of your computer; it is sometimes convenient to organize a workbook with multiple sheets, for instance sheets for data, analyses, bar graphs, or Visual Basic macros. Each sheet has a tab located at the bottom of the workbook (Fig I.8), and a sheet is activated when you click on its tab. Tab scrolling buttons allow you to navigate among the sheets. Clicking on a sheet tab on the bottom of the sheet activates it. Each workbook consists initially of three sheets labeled Sheet1, Sheet2, Sheet3 (16 initially in Excel 5). Sheets can be added, deleted, moved, and renamed to achieve a logical organization of data and analyses. To rename, move, delete, or copy a sheet, **right-click (Windows)** on a sheet tab or click and hold down the Control key on a **Macintosh**—we will refer to this as **Control-click**—and a pop-up menu appears from which you can select. A new sheet can also be inserted from the Menu Bar by choosing **Insert–Worksheet**. Note that a new worksheet is added to the left of the current or selected sheet. Sheets can also be deleted, copied, or edited from the Menu Bar using **Edit – Delete Sheet** or **Edit – Move or Copy Sheet. . . .**

Another way to move a sheet is to grab it by clicking on it and holding the mouse button (left button for Windows). A small icon of a paper sheet will appear under the mouse pointer. As you move the mouse pointer, you will notice a small dark marker moving between the sheet tabs. This marker indicates where the sheet will be moved when you release the mouse button.

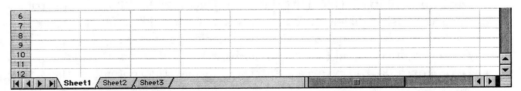

Figure I.8: Sheet Tabs

## I.4  Entering and Modifying Information

When a workbook is first opened, cell A1 automatically becomes the active cell. Active cells are surrounded by a dark outline indicating that they are ready to receive data. Use the mouse (or arrow keys on the keyboard) to activate a different cell. Then enter the data and either click on the Enter box or press the enter (return) key.

## Labels, Values, and Formulas

Three types of information can be entered into a cell: labels, values, and formulas. **Labels** are character strings such as words or phrases, typically used for headings or descriptions. They are not used in numerical calculations. **Values** are numbers such as 1.3, \$1.75, $\pi$. **Formulas** are mathematical expressions that use the values or formulas in other cells to create new values or formulas. All formulas begin with an equal (=) sign and are entered directly by hand in the cell or in the text entry area of the **Formula Bar** or by the **Function Wizard**. As an example of how a formula operates, if the formula

$$= A1 + A2 + A3 + A4$$

is entered in cell A5 and if the contents of cells A1, A2, A3, A4 are 11, 12, 19, $-6$, respectively, then cell A5 will show the value 36 because what is displayed in the cell is the result of the computation, not the formula. The formula in the cell may be viewed in the entry area in the **Formula Bar** if the cursor is placed over the cell.

It is the existence of formulas that makes a spreadsheet such a powerful tool. A formula such as

$$= \text{SUM}(A1 : A4)$$

is the Excel equivalent of the mathematical expression

$$\sum_{i=1}^{4} A_i$$

and a complex mathematical expression can be rendered into an Excel workbook in a similar fashion.

## Editing Information

There are several ways to edit information. If the data have not yet been entered after typing, then use the backspace or delete key or click on the red X to empty the contents of a cell. After the data have been entered, **activate** the cell by clicking on it. Then move the cursor to the text entry area of the Formula Bar where it turns into a vertical I-beam. Place the I-beam at the point you wish to edit and proceed to make changes.

## Cell References, Ranges, and Named Ranges

A **cell reference** such as A10 is a **relative** reference. When a formula containing the reference A10 is copied to another location, the cell address in the new location is changed to reflect the position of the new cell. For instance, if the formula presented earlier

$$= A1 + A2 + A3 + A4$$

that appears in cell A5 is copied to cell D9, then it will become

$$= D5 + D6 + D7 + D8$$

to reflect that the formula sums the values of the four cells just above its location.

This relative addressing feature makes it relatively easy to repeat a formula across a row or column of a sheet, such as adding consecutive rows, by entering the formula once in one cell, then copying its contents to the other cells of interest.

If you need to retain the actual column or row label when copying a formula, then precede the label with a dollar sign ($). This is called an **absolute** cell reference. For instance, $A2 keeps the reference to column A but the reference to row 2 is relative, A$2 leaves A as a relative reference, but fixes the row at 2, $A$2 gives the entire cell (row and column labels) an absolute reference. We will use mixed ($A2 or A$2) references in Chapter 9 with the chi-square distribution.

A group of cells forming a rectangular block is called a **range** and is denoted by something like A2:B4, which includes all the cells {A2, A3, A4, B2, B3, B4}. **Named ranges** are names given to individual cells or ranges. The main advantage of named ranges is that they make formulas more meaningful and easy to remember. We will use named ranges repeatedly in this book. To illustrate, suppose we wish to refer to the range A2:A7 by the name "data." Enter the label "data" in cell A1. Then select the range A1:A7 by clicking on A1, holding the mouse button down, dragging to cell A7, and then releasing the mouse button. From the Menu Bar choose **Insert − Name − Create**, and check the box **Top Row** then click **OK** (Fig I.9).

Perhaps the quickest way to add a named range is to select the cell(s) to be named and then click on the **Name box**. This creates the name and associates it with the selected cells.

You can see which names are in your workbook by clicking on the defined name drop-down list arrow to the right of the **Name box**. Names are displayed alphabetically.

If you type a name from your workbook in the **Name** box and hit Enter then you will be transported to the first entry in the corresponding range, which will appear in the text/formula entry area of the Formula Bar ready for editing. Named ranges are both an aid in remembering and constructing formulas and also a convenient way to move around your workbook. They are used extensively in this book.

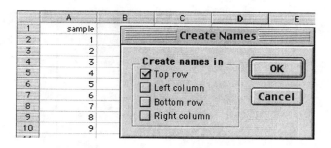

Figure I.9: Named Range

## Copying Information

To activate a block of cells, place the cursor in the upper left cell of the block, click and drag to the lower right cell and release. Alternatively activate the upper left cell, then press the **Shift** key and click on the lower right cell. Noncontiguous cells or blocks can be selected by holding down the **Command** key (**Macintosh**) or the **Control** key (**Windows**) while selecting each successive cell or block.

To copy data from a cell or range, activate it, then from the Menu Bar choose **Edit – Copy**. Move the mouse cursor to the new location and from the Menu Bar choose **Edit – Paste**.

An alternative is to activate the range, then place the mouse cursor on the border of the selected range. It will appear as a pointer. Now press the **Control** key and move the cursor to the location for copying. Release the mouse button.

To move data to another location choose **Edit – Cut** from the Menu Bar and then **Edit – Paste** after you move the mouse cursor to the new location. Alternatively, move the mouse cursor to the border of the selected range. It turns into a pointer. "Grab" the border with the mouse pointer and move the cells to the desired location.

Use of the mouse for copying and cutting is a **drag and drop** operation familiar to users of Microsoft Word.

## Shortcut Menu

If you activate a range and then **Control-click (Macintosh)** or **right-click (Windows)** on it, a **Shortcut Menu** will pop up next to the range allowing you access to some of the commands in the Menu Bar under **Edit**. This will provide options prior to copying or pasting.

## Paste Special

A useful command from the **Shortcut Menu** is the **Paste Special**. This provides a dialog box (Fig I.10) giving a number of options prior to pasting. The two options

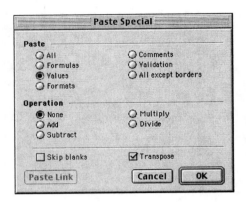

Figure I.10: Paste Special

most commonly used in this book are **transpose** check box, which transposes rows and columns, and the **Paste Values** radio button, which is useful if you need to copy a range of values defined by formulas. A straight copy will alter the cell references in the formulas and could produce nonsense. **Paste Special** solves this problem by pasting the values, not the formulas.

### Filling

Suppose you need to fill cells A1:A30 with the value 1. Enter the value 1 in cell A1. Activate A1 and move the cursor to the lower right-hand corner (the **fill handle**) of A1. The cursor becomes a cross hair. Drag the fill handle and pull down to cell A30. This copies the value 1 into cells A2:A30. Alternatively, you can select A1:A30 after entering the value 1 in cell A1. Then choose **Edit − Fill − Down** from the Menu Bar. Other options are available such as **Edit − Fill − Series** if this approach is taken.

## I.5   Opening Files

Often you will need to open text or data files, for instance the data files on the Student CD-ROM accompanying *PBS*. Other times, the data may be in a binary format produced by some other application.

Excel can read and open a wide selection of binary files. To see which ones may be imported, choose **File − Open** from the Menu Bar and make the appropriate selection from the drop-down list.

For text (ASCII) files, Excel may start the **Text Import Wizard**, after you make your selection from the drop-down list, once you choose **File − Open** from the Menu Bar. This is a sequence of three dialog boxes specifying how the text should be imported. The **Text Import Wizard** helps you make intelligent choices

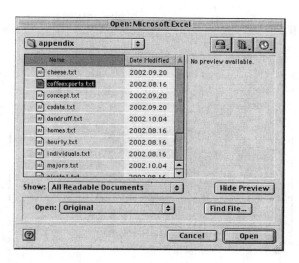

Figure I.11: Importing Files

and if the data file contains lines of explanatory text at the top of the file, then it is very handy for formatting correctly upon import. We illustrate using the "coffee export" data set on the Student CD-ROM. Fig I.11 shows that the file "coffeeexports.txt" has been selected following **File – Open – All Files**.

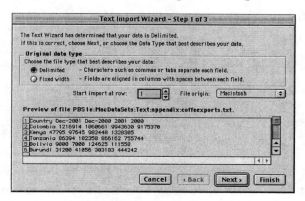

Figure I.12: Text Import Wizard—Step 1

Step 1. The Text Import Wizard (Fig I.12) makes a determination that the data were **Delimited**. You may override this with the radio button. Sometimes the Text Import Wizard interprets incorrectly. The default starting position is shown as Row 1. This too can be changed if labels are needed.

Step 2. The next screen depends on whether you chose Delimited or Fixed width in Step 1. If Delimited, you pick the delimiter. If Fixed width, you create line breaks.

Step 3.    The final step allows you to select how the imported data will be formatted. Usually the radio button **General**, the default, is appropriate.

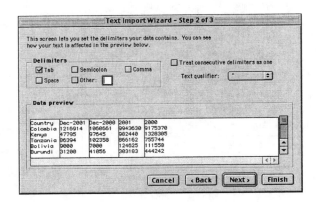

Figure I.13: Text Import Wizard—Step 2

## I.6    Printing

This is generally the last step. In the **Page Setup** dialog box (Fig I.14) accessed from the Menu Bar using **File − Page Setup**, there are four tabs:

- **Page.** Orientation, scaling, paper size, print quality.
- **Margins.** Top, bottom, left, right, preview window.
- **Header/Footer.** Information printed across top or bottom.
- **Sheet.** Print area, column/row title, print order.

The **Print Preview** button on the **Standard Toolbar** or **File − Print Preview** from the Menu Bar allows you to see what portion of your document is being printed and where it is positioned. Other options are available (Fig I.15) at the top of the print preview screen to assist you in previewing your output prior to printing.

Finally, when you are satisfied with your output, press the **Print...** button at the top of the preview screen. You can also print directly with the document window open using the **Print** button on the **Standard Toolbar** or **File − Print** from the Menu Bar. This brings up a **Print** dialog box in which you select which pages to print, the number of copies, printer setup, as well as buttons for **Page Setup** and **Print Preview** just discussed.

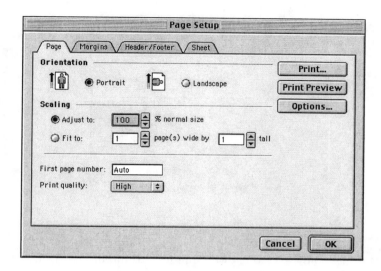

Figure I.14: Page Setup

Figure I.15: Print Preview Options

## I.7   Whither?

This brief introductory chapter contains a bare-bones description of Excel, as much as you need to know to access the remainder of this book. In the following chapters, not only will you make use of many of the topics and tips presented here, but you will learn about the **ChartWizard**, enter formulas, copy and paste cells, and so on. By using Excel for statistics you will also obtain a good practical background in spreadsheets.

One way to learn more about Excel is by using the animated **Office Assistant** as in Fig I.16. You can install a gallery of different characters from the Excel installation CD-ROM in the Office:Actors folder. The one shown here is called **Max**. Max can be called up using the **Help** button (the ? at the right of the **Standard Toolbar**). Ask Max a question such as "What's new?" and see how he responds. If you **right-click (Windows)** or **Control-click (Macintosh)** Max, then more options become available.

Numerous books are available in libraries and in bookstores, but books often contain too much information. Finally, there is a wealth of recent material available on the Internet. Use your favorite search engine or directory and you'll find pointers to macros (Visual Basic programs) and sample workbooks made available by other

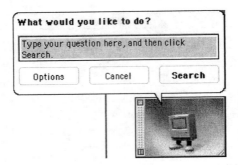

Figure I.16: Office Assistant—Max

users for applications of Excel. Current versions of Excel are "Internet ready" and can read html files and carry out "web queries" to import data directly from the Internet or save Excel files to html format. In fact, with some browsers and servers it is also possible for users to manipulate spreadsheets within the browser.

If you find interesting Internet resources, let me know (*hoppe@mcmaster.ca*) and I will place a pointer to them on the Excel statistics Web site, which will be located at http://www.mathematics.net/.

# Chapter 1

# Looking at Data–Distributions

Excel provides more than 70 functions related to statistics and data analysis as well as tools in the **Analysis ToolPak**. Additionally, the **ChartWizard** gives a step-by-step approach to creating informative graphs.

We first discuss the **ChartWizard**. The figures are from Excel 2001 for **Macintosh**. They differ only cosmetically from Excel 2000 for **Windows**, as illustrated in Figure 1.1.

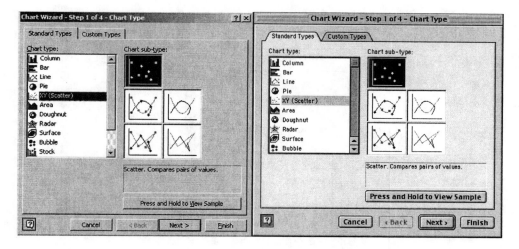

Figure 1.1: Windows 2000 and Macintosh 2001 ChartWizard Opening Screens

## 1.1   Displaying Distributions with Graphs

### The ChartWizard

The **ChartWizard** is a step-by-step interface for creating informative graphs (called Charts in Excel). There are four steps in **Excel 97/2000 (Windows)** and **Excel 98/2001 (Macintosh)** that guide the user by requesting details about the chart type, formatting, titles, legends, and so on, in dialog boxes. The ChartWizard can be activated either from the button on the **Standard Toolbar** or by choosing **Insert – Chart** from the Menu Bar.

> **Example 1.1.**   (Page 6 in text.) In the summer of 2000, Firestone received much attention in the media due to a number of traffic accidents believed to be caused by tread separation in Firestone tires. Figure 1.2 shows the distribution of the counts for tire models for 2969 accidents. Construct a bar chart of this data.

| | A | B | C |
|---|---|---|---|
| 1 | **Tire Model** | **Count** | **Percent** |
| 2 | ATX | 554 | 18.7 |
| 3 | Firehawk | 38 | 1.3 |
| 4 | Firestone | 29 | 1.0 |
| 5 | Firestone ATX | 106 | 3.6 |
| 6 | Firestone Wilderness | 131 | 4.4 |
| 7 | Radial ATX | 48 | 1.6 |
| 8 | Wilderness | 1246 | 42.0 |
| 9 | Wilderness AT | 709 | 23.9 |
| 10 | Wilderness HT | 108 | 3.6 |

Figure 1.2: Firestone Tire Data

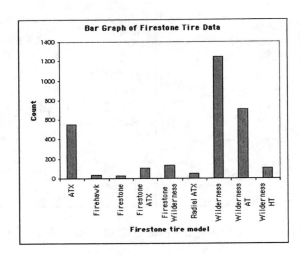

Figure 1.3: Excel Bar Chart

**Solution.** Figure 1.3 is a bar graph produced by Excel that displays the same information as in Figure 1.2. For other types of graphical displays, make appropriate choices from the same sequence of dialog boxes. The following steps describe how it is obtained. First, enter the data and labels in cells A1:C10. Note that we have included the column "Percent" in order to explain how to select or eliminate columns of data from a table.

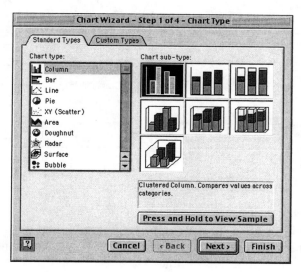

Figure 1.4: ChartWizard—Step 1

Step 1. Select cells A1:C10 and click the **ChartWizard**. The ChartWizard (Figure 1.4) displays the types of graphs that are available. In the left field select **Column** for Chart type. In the right field select **Clustered Column** for Chart sub-type, which is the first choice in the top row. When you select a sub-type, an explanation of the chart appears in the box below all the choices, and you can preview your chart's appearance using the **Press and Hold to View Sample** button. Click **Next**.

Step 2. The next dialog box (Figure 1.5) with title **Chart Source Data** previews your chart and allows you to select the data range for your chart. Since you already selected cells A1:C10 prior to invoking the ChartWizard, this block appears in the text area **Data range**. Otherwise you would input the range now, or you can make corrections to the data range. The preview chart shows bar charts for both the counts and the percents. To remove the percents, click the **Series** tab at the top of the dialog box, then highlight the series "Percent" and click the **Remove** button (right side of Figure 1.5). The bar graph of percents vanishes. **Click Next.**

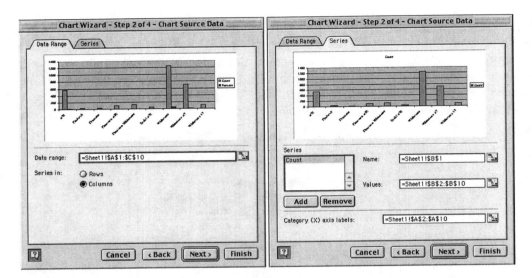

Figure 1.5: ChartWizard—Step 2

**Step 3.** A dialog box **Chart Options** (Figure 1.6) appears with the default chart. Rarely is the default satisfactory; you will generally need to make cosmetic changes to its appearance.

- Click the **Titles** tab. Enter "Bar Graph of Firestone Tire Data" for Category (X) axis and "Count" for Value (Y) axis.

- Click the **Legend** tab. We don't require a legend because only one variable is plotted, so make sure the check box **Show Legend** is cleared.

- Additional tabs are available to customize other types of charts. They are not required here. Click **Next**.

**Step 4.** The final step lets you decide if you want the chart placed on the same worksheet as the data or on another worksheet. With each choice there is a field for entering the worksheet name. We will embed the chart on the same worksheet, so we select the radio button **As object in:** (Figure 1.7). As our current workbook only has one sheet, Excel has used the default name Sheet1. We could also embed the chart on another sheet in the same workbook. Click the **Finish** button.

The chart appears with eight handles indicating that it is selected. The chart can be resized by selecting a handle and then dragging the handle to the desired size. The chart can also be moved. Click the interior of the chart and drag it to another location (holding the mouse button down). Then click outside the chart to deselect.

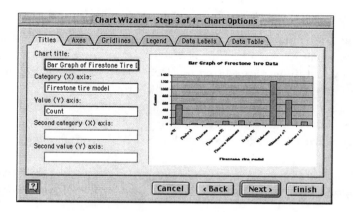

Figure 1.6: ChartWizard—Step 3

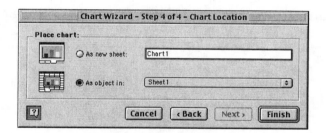

Figure 1.7: ChartWizard—Step 4

You will also find the **Chart Toolbar** (Figure 1.8) embedded on the worksheet. This is used for embellishments of the chart. Use of this toolbar is described in the next section on creating histograms. Note that the Chart Toolbar may also be called from the Menu Bar by **View – Toolbars – Chart**. Also, if you select the Chart by clicking once within its area, the Menu Bar will change. In place of the word **Data**, there will now appear **Chart** from which a pull-down menu will provide the same tools as are displayed with icons on the Chart Toolbar.

**Note:** You might get an error message "Cannot add chart to shared workbook" even if you are not sharing your workbook. This is a bug introduced when Shared Workbooks were implemented in Excel 97/98. It occurs under certain conditions

Figure 1.8: Chart Toolbar

if you try to create a chart using the data analysis tools.

   You can still output your chart to a new workbook and then, if desired, copy the chart to the existing workbook. This is one workaround. Fortunately, this problem can be fixed by installing an updated file **ProcDBRes** to replace the existing one of the same name in the folder/directory "Microsoft Office 98:Office:Excel Add-Ins:Analysis Tools" (the folder for a default installation—your location may differ). This file is available for download at the Microsoft Software Library. There was a similar problem with Excel 97 (fixed in Excel 2000). Details may be found at the two URLs

`http://support.microsoft.com/default.aspx?scid=kb;EN-US;q183188`
`http://support.microsoft.com/default.aspx?scid=kb;en-us;Q178243.`

## Pie Charts

It is easy as pie to produce a chart as in Figure 1.9 using Excel. Select **Pie** in place of **Column** in Step 1 of the ChartWizard and follow the remaining steps.

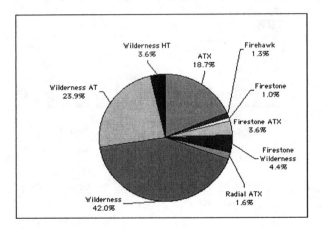

Figure 1.9: Pie Chart

   Alternatively, since we have already created a bar graph, it is instructive to use the **Chart Toolbar** as an illustration of how easily modifications may be made. This interface is a vast improvement over the previous version of Excel.

1. Select the completed bar graph by clicking once within its border.

2. From the Menu Bar select **Chart – Chart Type. . . .** You will be presented with a box that is identical to Figure 1.4 but for the title, which contains only the words **Chart Type** without mention of Step 1 of the ChartWizard. In the left field, referring to Figure 1.4, select **Pie** for Chart type and in the right field select **Pie** for Chart sub-type, the first choice in the top row.

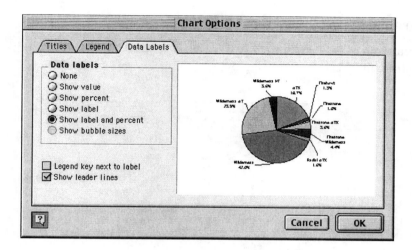

Figure 1.10: Chart Options

3. From the Menu Bar select **Chart − Chart Options....** Figure 1.10 appears with three tabs: Titles, Legend, and Data Labels. Click the tab **Data Labels** and then select the radio button **Show label and percent** and check the box **Show leader lines.** Click **OK**. A pie chart now replaces the bar graph.

4. Double-clicking on any label will bring up the **Format Data Label** dialog box, which allows you to change background colors, number of decimals, and other formatting features.

## Histograms

The ChartWizard is designed for use with data that are already grouped, for instance, categorical variables, and can therefore be used to construct a histogram of quantitative variables that have been grouped into categories or intervals. For raw numerical data, Excel provides additional commands within the **Analysis ToolPak** for constructing histograms.

To determine whether this toolpak is installed, choose **Tools − Add-Ins** from the Menu Bar. The **Add-Ins** dialog box appears. Depending on whether other Add-Ins have been loaded, your box might appear slightly different. If the **Analysis ToolPak** box is not checked, then select it and click **OK**. The Analysis ToolPak will now be an option in the pull-down menu when you choose **Tools − Data Analysis.** Note that you can also use the **Select** button to add customized add-ins to complement Excel.

## Histogram from Raw Data

**Example 1.2.** (Example 1.3, Table 1.1, page 11 in text.) Make a histogram of the state unemployment rates in December 2000.

| | A | B | C | D |
|---|---|---|---|---|
| 1 | Unemployment Rates by State, December 2000 | | | |
| 2 | 4 | 5.3 | 3.7 | |
| 3 | 6.1 | 2.6 | 2.6 | |
| 4 | 3.3 | 3.3 | 4 | |
| 5 | 3.9 | 2 | 3.8 | |
| 6 | 4.3 | 3.4 | 8.9 | |
| 7 | 2.1 | 2.8 | 3.2 | |
| 8 | 1.5 | 4.3 | 3.3 | |
| 9 | 3.3 | 3.2 | 2.3 | |
| 10 | 3.2 | 4.9 | 3.8 | |
| 11 | 3 | 2.5 | 3.4 | |
| 12 | 3.6 | 4 | 2.7 | |
| 13 | 5 | 2.2 | 2.4 | |
| 14 | 4.5 | 3.5 | 1.9 | |
| 15 | 2.7 | 4.9 | 4.9 | |
| 16 | 2.5 | 4.2 | 5.5 | |
| 17 | 3.2 | 3.6 | 3 | |
| 18 | 3.7 | 2.7 | 3.7 | |

Figure 1.11: Unemployment Rates by State, December 2000

**Solution.** Excel requires a contiguous block of data for the histogram tool. We have entered the data in block A2:C18 in Figure 1.11.

1. From the Menu Bar choose **Tools – Data Analysis** and scroll to the choice **Histogram** (Figure 1.12). Click **OK**.

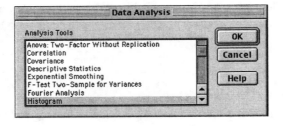

Figure 1.12: Data Analysis Tools

2. In the dialog box (Figure 1.13) type the reference for the range A2:C18 in the **Input range** area, which is the location on the workbook for the data. As with the Bar Chart, you may instead click and drag from cell A2 to C18. The choice depends on whether your preference is for strokes (keyboard) or clicks (mouse). Leave the **Bin range** blank to allow Excel to select the bins, check the **Labels** box blank, type a cell location, say E1, for **Output range** to denote the upper left cell of the output range, and

check the box **Chart output** to obtain a histogram on the same sheet of the workbook as the data. The option **Pareto** (sorted histogram) constructs a histogram with the vertical bars sorted from left to right in decreasing height. If **Cumulative Percentage** is checked, the output will include a column of cumulative percentages.

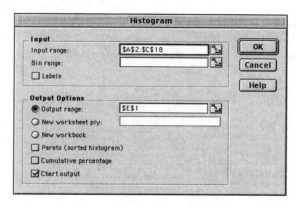

Figure 1.13: Histogram Tool

3. The output appears in Figure 1.14. The entries under "Bin" in E2:E6 are not the midpoints of the bin intervals, as you might expect. Rather they are the **upper limits** of the boundaries for each interval. The corresponding frequencies appear in cells F2:F6 with the histogram to the right. We shall shortly modify the histogram by changing the labels and allowing adjacent bars to touch. But first, we explain how to customize the selection of bins.

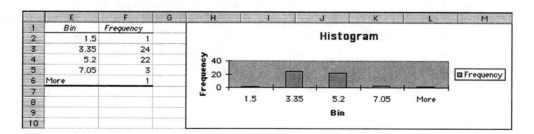

Figure 1.14: Output Table and Default Histogram

## Changing the Bin Intervals

If the bin intervals are not specified, then Excel creates them automatically, choosing the number of bins roughly equal to the square root of the number of observations, beginning and ending at the minimum and maximum, respectively, of the data set. In creating the above histogram from raw data, we allowed Excel to choose the default bins. Here, we shall redo the histogram by selecting our own bin intervals.

1. Type "Bin" (or another appropriate label) in an empty location, say E1. Then enter the values 0.9, 1.9, 2.9, 3.9, 4.9, 5.9, 6.9, 7.9, 8.9 directly below in cells E2:E10. An easy way to accomplish this is to type 0.9 and 1.9 in cells E2 and E3 respectively, then select E2 and E3, click the fill handle in the lower right corner of E3, then drag the fill handle down to cell E10 and release the mouse button.

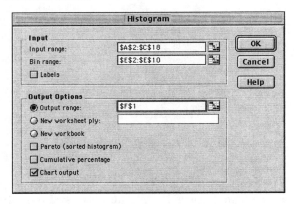

Figure 1.15: Histogram Tool—Specified Bin Intervals

2. Repeat the earlier procedure for creating a histogram, but this time type E2:E10 in the text area for **Bin range** and select a location (cell F1) in Figure 1.15) to mark where the output with your selected bin intervals will appear (see Figure 1.16). Excel echos the selected bin intervals in F2:F10 as part of its output and also inserts "More" in F11 for any data beyond the bin range.

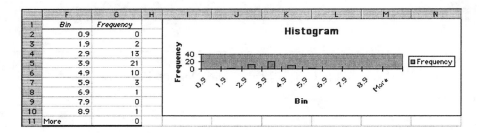

Figure 1.16: Output Table and New Bin Histogram

### Enhancing the Histogram

Although the default histogram captures the overall features of the data set, it is inadequate for presentation. Excel provides a set of tools for enhancing the histogram. These are too numerous for all to be mentioned here, but a few will be discussed with reference to the example. The other options may be invoked analogously.

**Legend.** To remove the legend (which is not needed here) select **Chart − Chart Options...** from the Menu Bar, click the **Legend** tab, and clear the box **Show legend**.

**Resize.** Both the histogram (called the **Plot Area**) and the box that contains it (called the **Chart Area**) can be resized and moved. Select the Chart Area by clicking once within its boundary, resize using any of the eight handles that appear, or move it by dragging, or cutting and pasting from **Edit** on the Menu Bar, to a new location. Likewise, select the Plot Area by clicking once within its boundary and then resize or move as with the Chart Area. The X axis labels may appear placed by default horizontally, vertically, or diagonally to accommodate the selected size. This can also be changed. After removing the Legend and resizing, click outside the Chart Area to deselect.

**Bar Width.** Adjacent bars do not touch in the default histogram, which looks more like a bar chart for categorical data. To adjust the bar width, click and select any one of the bars, and then from the Menu Bar select **Format − Selected Data Series...** to bring up the **Format Data Series** dialog box. Select the **Options** tab (Figure 1.17) and change the **Gap width** from 150% to 0%.

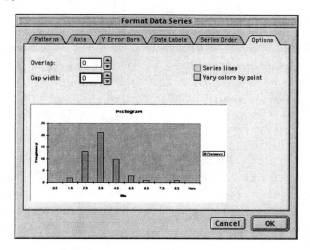

Figure 1.17: Format Data Series

**Chart Title.** Click on the title word "Histogram". A rectangular grey border with handles will surround the word, indicating that it is selected for editing. Begin typing "Unemployment Rates by State", then hold down the **Alt** key (**Windows**) or the **Command** key (**Macintosh**), and press **Enter**. You may now type a second line of text in the **Formula Bar** entry area. Continue typing "December 2000" and then press **Enter**. If you want to move the title within the Chart Area, use the handles. To change the font of your title, select the title, then from the Menu Bar choose **Format – Selected Chart Title....** The dialog box has three tabs: **Patterns, Font,** and **Alignment**. Select the **Font** tab and pick a font face, style, and size.

**X axis Title.** Click on the word **Bin** at the bottom of the chart and then type "Unemployment rate". Change the font by selecting the X axis title, then from the Menu Bar choose **Format – Selected Axis Title...** and complete the dialog box as desired in the same fashion as for the Chart Title.

**Y axis Title.** Click on the word **Frequency** on the left side, and then from the Menu Bar choose **Format – Selected Axis Title...** for any desired formatting.

**X axis Format.** Double-click the X axis, and in the **Format Axis** dialog box you can click on various tabs to change the appearance of the X axis. If you click on the **Alignment** tab, you can change the orientation of the X axis labels.

**Y axis Format.** Double-click the Y axis, and in the **Format Axis** dialog box, you can click on various tabs to change the appearance of the Y axis. Sometimes, if you resize the chart, you will need to experiment with the **Major** or **Minor** units to achieve a pleasing result. Click **OK**.

**More Interval.** The "More" interval with 0 counts is a nuisance. Delete this bin and the label will disappear from the plot. However, if there are counts in this bin, you can change the label "More" to a numerical value and, because the histogram is dynamically linked to the frequency table, this change will also appear on the histogram. In Figure 1.18 we changed "More" to the value "10".

**Plot Area Pattern.** The default histogram has a border around the Plot Area and the Plot Area is shaded grey. Both defaults can be changed by double-clicking the **Plot Area** to bring up the **Format Plot Area** dialog box. If desired, select the radio button **None** for Border and also select the radio button **None** for Area.

**Bin End Points.** The values $0.9, 1.9, \ldots, 9.9$ are unattractive. Change them to $1.0, 2.0, \ldots, 10$ as in Figure 1.18.

At the conclusion of the formatting, the histogram will look like Figure 1.18.

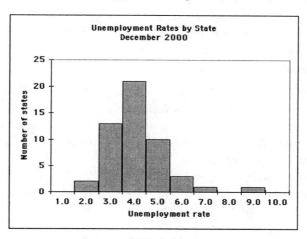

Figure 1.18: Final Histogram after Editing

## Three-dimensional Histogram

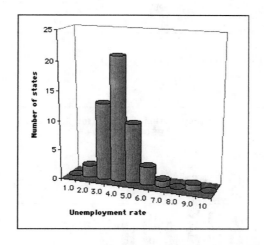

Figure 1.19: Three-dimensional Histogram

A nice feature of Excel is the ease of convertiing a chart into any other type. This is especially useful in preparing output for presentation. Figure 1.19 shows a three-dimensional histogram obtained from Figure 1.18. Select the previous histogram and from the Menu Bar choose **Chart – Chart Type ....** For Chart type select **Cylinder** and for Chart sub-type, select the first (upper left) choice.

## Histogram from Grouped Data

The **Histogram** tool requires the raw data as input. When numerical data have already been grouped into a frequency table, the **ChartWizard** is the appropriate tool. First, use it to obtain a bar chart, and then modify it exactly as you would enhance a histogram.

**Example 1.3.**    (See Exercise 1.18, page 26 in text.)   Figure 1.20 gives the age distribution of US residents in 1950 and 2075 in millions of people.  Because the total population in 2075 is much larger than the 1950 population, comparing relative frequencies in each group is clearer than comparing counts. Make a relative frequency histogram of the 1950 age distribution and the projected 2075 age distribution, on the same graph and with the same scale. The final histogram should be similar to that shown in Figure 1.21.

| | A | B | C | D |
|---|---|---|---|---|
| 1 | Age group | Bin | 1950 | 2050 |
| 2 | under 10 | 10 | 29.3 | 53.3 |
| 3 | 10 – 19 | 20 | 21.8 | 53.2 |
| 4 | 20 – 29 | 30 | 24 | 51.2 |
| 5 | 30 – 39 | 40 | 22.8 | 50.5 |
| 6 | 40 – 49 | 50 | 19.3 | 47.5 |
| 7 | 50 – 59 | 60 | 15.5 | 44.8 |
| 8 | 60 – 69 | 70 | 11 | 40.7 |
| 9 | 70 – 79 | 80 | 5.5 | 30.9 |
| 10 | 80 – 89 | 90 | 1.6 | 21.7 |
| 11 | 90 – 99 | 100 | 0.1 | 8.8 |
| 12 | 100 – 109 | 110 | 0 | 1.1 |

Figure 1.20: Age Distribution - Grouped Data

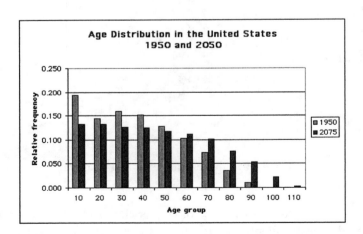

Figure 1.21: Histogram from Grouped Data

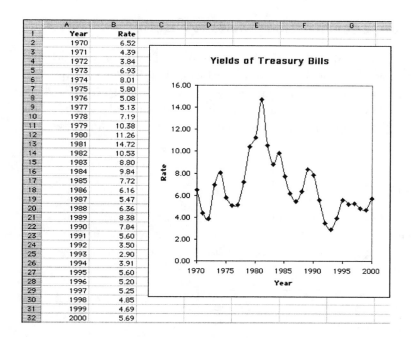

| | A | B | C | D | E | F | G |
|---|---|---|---|---|---|---|---|
| 1 | Year | Rate | | | | | |
| 2 | 1970 | 6.52 | | | | | |
| 3 | 1971 | 4.39 | | | | | |
| 4 | 1972 | 3.84 | | | | | |
| 5 | 1973 | 6.93 | | | | | |
| 6 | 1974 | 8.01 | | | | | |
| 7 | 1975 | 5.80 | | | | | |
| 8 | 1976 | 5.08 | | | | | |
| 9 | 1977 | 5.13 | | | | | |
| 10 | 1978 | 7.19 | | | | | |
| 11 | 1979 | 10.38 | | | | | |
| 12 | 1980 | 11.26 | | | | | |
| 13 | 1981 | 14.72 | | | | | |
| 14 | 1982 | 10.53 | | | | | |
| 15 | 1983 | 8.80 | | | | | |
| 16 | 1984 | 9.84 | | | | | |
| 17 | 1985 | 7.72 | | | | | |
| 18 | 1986 | 6.16 | | | | | |
| 19 | 1987 | 5.47 | | | | | |
| 20 | 1988 | 6.36 | | | | | |
| 21 | 1989 | 8.38 | | | | | |
| 22 | 1990 | 7.84 | | | | | |
| 23 | 1991 | 5.60 | | | | | |
| 24 | 1992 | 3.50 | | | | | |
| 25 | 1993 | 2.90 | | | | | |
| 26 | 1994 | 3.91 | | | | | |
| 27 | 1995 | 5.60 | | | | | |
| 28 | 1996 | 5.20 | | | | | |
| 29 | 1997 | 5.25 | | | | | |
| 30 | 1998 | 4.85 | | | | | |
| 31 | 1999 | 4.69 | | | | | |
| 32 | 2000 | 5.69 | | | | | |

Figure 1.22: Yield of Treasury Bills

## Time Plot

A Time Plot of a variable plots each observation against the time at which it was measured. Time goes on the horizontal axis and the variable being measured goes on the vertical axis. Connecting the data points by lines helps emphasize any change over time.

> **Example 1.4.** (See Exercise 1.12, page 23 in text.) Treasury bills are short-term borrowing by the U.S. government. Figure 1.22 gives the annual returns on Treasury bills from 1970 to 2000. Make a Time Plot of the returns.

**Solution.** Select the block A1:B32 and invoke the **ChartWizard**. Select Chart type **XY (Scatter)** and the second Chart sub-type (Scatter with data points connected by smoothed Lines.). Under **Chart Options** deselect all gridlines and label the axes. Format the X-axis so the minimum is 1970 and the maximum is 2000. The output appears in Figure 1.22.

## 1.2   Describing Distributions with Numbers

The most direct way to obtain the common summary statistics is through the **Descriptive Statistics Tool**, which provides preformatted output very quickly. It is explained in this section. An alternative is the **Formula Palette**, which provides greater flexibility of output, and many more functions and formulas over its predecessor. We first describe the Descriptive Statistics Tool and then the Formula Palette.

### The Descriptive Statistics Tool

**Example 1.5.** (See Exercise 1.30, page 34 in text.)   This exercise gives the amounts spent by 50 consecutive shoppers at a supermarket. Describe the data using the Descriptive Statistics Tool.

| | A | B | C | D |
|---|---|---|---|---|
| 1 | **Supermarket Spending** | | *Supermarket Spending* | |
| 2 | 3.11 | | | |
| 3 | 8.88 | | Mean | 34.7022 |
| 4 | 9.26 | | Standard Error | 3.068 |
| 5 | 10.81 | | Median | 27.855 |
| 6 | 12.69 | | Mode | #N/A |
| 7 | 13.78 | | Standard Deviation | 21.697 |
| 8 | 15.23 | | Sample Variance | 470.777 |
| 9 | 15.62 | | Kurtosis | 0.709 |
| 10 | 17.00 | | Skewness | 1.103 |
| 11 | 17.39 | | Range | 90.23 |
| 12 | 18.36 | | Minimum | 3.11 |
| 13 | 18.43 | | Maximum | 93.34 |
| 14 | 19.27 | | Sum | 1735.11 |
| 15 | 19.50 | | Count | 50 |
| 16 | 19.54 | | Largest(5) | 70.32 |
| 17 | 20.16 | | Smallest(5) | 12.69 |
| 18 | 20.59 | | | |

Figure 1.23: Descriptive Statistics Output

**Solution.**   Enter the data in block A2:A51 of a workbook with the label in A1.

1. From the Menu Bar choose **Tools − Data Analysis** and then double-click **Descriptive Statistics** (or, equivalently, select **Descriptive Statistics** and click **OK**) in the **Data Analysis Dialog** box. A dialog box **Descriptive Statistics** appears, which prompts for user input.

2. Complete the input as follows. The **Input range:** is A1:A51. If you selected this range prior to invoking **Descriptive Statistics**, it will already be inserted by Excel. Check the box **Labels in first row**. The **Confidence level for mean:** is not needed at this time (it gives the half-width). Check the **Kth largest:** or **Kth smallest:** boxes if needed. We have selected K = 5 for illustration.

3. The **Output Options** tell Excel where to place the output. Select cell C1. Finally, check the box **Summary Statistics** and click **OK**. The output appears in Figure 1.23. We have formatted the output by reducing the number of decimal points using the **Decimal** button in the **Formatting Toolbar**.

## Formula Palette

The **Function Wizard** was replaced by the **Formula Palette** in Excel 97/98. This is a tool that assists in entering formulas and functions included in Excel, particularly complex ones. The functions can perform decision-making, action-taking, or value-returning operations. The Formula Palette simplifies this process by guiding you step by step.

It can be fired up in one of two ways. When you select a cell and press the

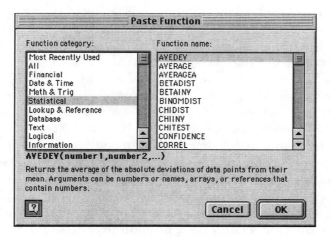

Figure 1.24: Paste Function

**Paste Function** button $f_x$ next to the autosum button $\Sigma$ on the **Standard Toolbar** (or, equivalently, choose **Insert − Function...** from the Menu Bar), an equal sign (=) appears both in the cell and in the **Formula Bar**. The **Paste Function** dialog box (Figure 1.24) appears showing all available functions grouped by category on the left and the function name on the right. Both lists have scroll

Figure 1.25: Formula Palette—Default

bars for choices not directly visible on the screen. At the bottom of the box, the selected function is shown, with the arguments it takes and a brief description. (In previous versions of Excel, a similar dialog box called **Function Wizard – Step 1 of 2** appeared.) When you click **OK** in the **Paste Function** box, the **Formula Palette** box appears below the **Formula Bar**, requesting parameters and an input range for the function you have selected. In addition, the **Formula Bar** is now activated, showing the **Formula Palette's** drop-down list control, with the 10 most recently used functions. An equal (=) sign appears in the **Formula Bar**, showing the selected function partially constructed and awaiting completion of its arguments. You may enter these either directly into the **Formula Bar** or in the **Formula Palette** box.

The **Formula Palette** is usually invoked in a second, more direct way. Select a cell and press (=) the **Formula Bar** to open the **Formula Palette** dialog box (Figure 1.25). On the far left side of the **Formula Toolbar** is a button with the most recently used function, in this case SUM. If this is the function you need, then click on the word SUM, and the **Formula Palette** dialog box will expand, requesting the required parameter or the data range for the function. (This can be typed directly or *referenced*, by using the mouse to point to the data by clicking and dragging over cells in the data range.) As you input this information, Excel will correspondingly build the function, both in the **Formula Bar** and in the cell you had selected in the workbook. When you have completed entering the requested input, click **OK** to complete the function. If you want some function other than the default, click the small arrow in the box to the right of the function name to generate a drop-down list of your 10 most recently used functions or you may select **More functions. . . .** If you select the latter, then the **Paste Function** dialog box appears. An OK (checkmark symbol) and a cancel (X) button appear to the right of this arrow. Click the checkmark, and the formula is entered into the active cell. Click the cancel to discard the formula without making changes.

**Recommendation.** The **Paste Function** button on the Standard Toolbar duplicates the actions of the **Formula Palette**. Because Excel formulas start with an equal (=) sign, we recommend that you begin your formulas by pressing the (=) symbol on the Formula Toolbar instead of using the Paste Function. This activates the **Formula Palette**, and you can either type the formula by hand into the Formula Bar or order up a function from the **Paste Function** box, if required. Experienced users of Excel often customize the **Standard Toolbar** and replace the Paste Function button with some other one.

## The Five-Number Summary

**Example 1.6.**    Find the five-number summary {minimum, first quartile, median, third quartile, maximum} for the supermarket spending data.

| | A | B | F | G | H |
|---|---|---|---|---|---|
| 1 | Supermarket Spending | | | Five-number summary | |
| 2 | 3.11 | | | | |
| 3 | 8.88 | | Minimum | =MIN(A2:A51) | 3.11 |
| 4 | 9.26 | | First quartile | =QUARTILE(A2:A51,1) | 19.3275 |
| 5 | 10.81 | | Median | =MEDIAN(A2:A51) | 27.86 |
| 6 | 12.69 | | Third quartile | =QUARTILE(A4:A23,1) | 15.5225 |
| 7 | 13.78 | | Maximum | =MAX(A2:A51) | 93.34 |

Figure 1.26: Five-Number Summary

**Solution**

1. Refer to Figure 1.26. Enter the labels "Minimum," "First quartile," "Median," "Third quartile," and "Maximum" in F3:F7, as shown.

2. Select cell H3 and then click the equal (=) symbol on the **Formula Bar** to start the **Formula Palette**, and use the drop-down list to select **More functions....** In the **Paste Function** dialog box, select **Statistical** from the left and scroll down to select QUARTILE on the right. Click **OK**.

3. The **Formula Palette** dialog box appears. Move it out of the way and enter the data **Array** by selecting cells A2:A51 with your mouse (or more mundanely, by typing A2:A51 into the dialog box). Click in the text area for **Quart** and type "1" to indicate the first quartile. The completed formula appears in the **Formula Toolbar**; the value of the formula 19.3275 shows in the dialog box (Figure 1.27). Click **OK**. The value 19.3275 is printed in H5.

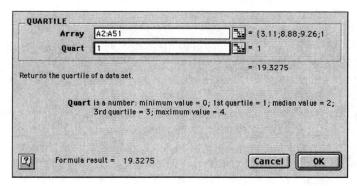

Figure 1.27: Quartile Formula

4. Continue in this fashion, using the **Formula Palette** to complete the five number summary. Of course, you can still enter the formulas by hand in the **Formula Bar** once you are familiar with them. Cells G3:G7 show the syntax, while the values are in H3:H7.

The five-number summary is {3.11, 19.33, 27.86, 115.52, 93.34}. We observe that Excel uses a slightly different definition of quartiles for a finite data set than the text.

Next we discuss the boxplot macro, which also displays the five-number summary.

## Macro – Boxplot

| | A | B | C | D |
|---|---|---|---|---|
| 1 | Annual Earnings of Hourly Workers at National Bank | | | |
| 2 | Black females | Black males | White females | White males |
| 3 | 16015 | 18365 | 25249 | 15100 |
| 4 | 17516 | 17755 | 19029 | 22346 |
| 5 | 17274 | 16890 | 17233 | 22049 |
| 6 | 16555 | 17147 | 26606 | 26970 |
| 7 | 20788 | 18402 | 28346 | 16411 |
| 8 | 19312 | 20972 | 31176 | 19268 |
| 9 | 17124 | 24750 | 18863 | 28336 |
| 10 | 18405 | 16576 | 15904 | 19007 |
| 11 | 19090 | 16853 | 22477 | 22078 |
| 12 | 12641 | 21565 | 19102 | 19977 |
| 13 | 17813 | 29347 | 18002 | 17194 |
| 14 | 18206 | 19028 | 21596 | 30383 |
| 15 | 19338 | | 26885 | 18364 |
| 16 | 15953 | | 24780 | 18245 |
| 17 | 16904 | | 14698 | 23531 |
| 18 | | | 19308 | |
| 19 | | | 17576 | |
| 20 | | | 24497 | |
| 21 | | | 20612 | |
| 22 | | | 17757 | |

Figure 1.28: Salaries Data for Boxplot

A boxplot is one of the most important exploratory tools available to the data analyst. Unfortunately, this tool is not part of the Analysis ToolPak. The Microsoft Personal Support Center does have a Web page "XL: How to Create a BoxPlot – Box and Whisker Chart (Q155130)" located at

http://support.microsoft.com/support/kb/articles/q155/1/30.asp

with instructions for creating a reasonable boxplot, using a **Volume-Open-High-Low-Close** Chart, but the sequence of steps is complicated enough to discourage a student from using this tool regularly. We have provided a macro *BP.xls* which will produce a boxplot of a single data set or side-by-side boxplots of multiple data sets.

> **Example 1.7.** (See Case 1.2, page 32 in text.) Banks employ many workers paid by the hour, such as tellers and data clerks. National Bank has been accused of discrimination in paying its hourly workers. Figure 1.28 shows the annual salaries for a sample of workers (Table 1.8, page 32 in the text). Use the boxplot macro to produce side-by-side boxplots.

**Solution.**    There are two ways to activate the macro, depending on where the data are located.    In either case, the Excel file BP.xls containing the boxplot macro, needs to be open.  When you open this file you will be prompted with a warning (Figure 1.29) that the workbook contains a macro.  This is normal; you should select "Enable Macros."  To use this macro, the data must be in columns

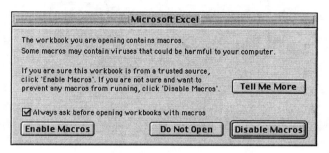

Figure 1.29: Macro Warning

on a worksheet.  However, the data need not be in adjacent columns nor need the number of observations in each data set be the same.  If the data sets have the same column lengths, then an entire block can be selected.  Otherwise, each column needs to be selected separately, as will be discussed in this example.

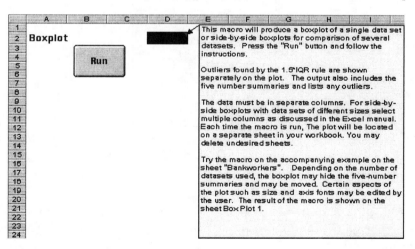

Figure 1.30: Boxplot Macro

1. When the data are on the same worksheet as the boxplot macro, then click the button **Run Boxplot** (see Figure 1.30) to bring up the Boxplot dialog box (Figure 1.32).

2. When the data are in another workbook then the macro is run as follows. From the Menu Bar select **Tools – Macro – Macros ...** to open the **Macro** dialog box (Figure 1.31 which lists all available macros. 1.31. Highlight

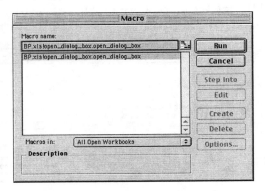

Figure 1.31: Boxplot Macro Select Box

BP.xls and then select **Run**, which will open Figure 1.32.

Figure 1.32: Boxplot Dialog Box

3. You now need to select the data. In this example, first select cells A2:A17, then hold down the **Control (Windows)** or **Command (Macintosh)** keys and select B2:B17, C2:C17, and D2:D17. Check the box **First Row Contains Labels.** You can also label the data axis (X-axis) if desired. The boxplot, Figure 1.32, will appear in a new sheet named "Box Plot 1" in your workbook, together with the five-number summaries for the data sets. Outliers are also plotted.

Figure **1.33** shows the resulting boxplot together with the five-number summary. Outlier values are also printed according to the $1.5 \times IQR$ rule.

Because Excel uses a slightly different definition for the quartiles than is given in the text, there may be occasional differences between the Excel output and what is given in the text. This may affect judgment on which observations would be considered outliers.

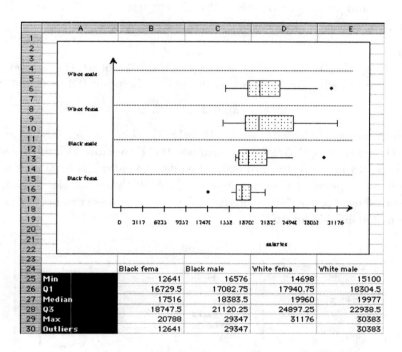

Figure 1.33: Boxplot Output

| | | Black fema | Black male | White fema | White male |
|---|---|---|---|---|---|
| 25 | Min | 12641 | 16576 | 14698 | 15100 |
| 26 | Q1 | 16729.5 | 17082.75 | 17940.75 | 18304.5 |
| 27 | Median | 17516 | 18383.5 | 19960 | 19977 |
| 28 | Q3 | 18747.5 | 21120.25 | 24897.25 | 22938.5 |
| 29 | Max | 20788 | 29347 | 31176 | 30383 |
| 30 | Outliers | 12641 | 29347 | | 30383 |

## 1.3   The Normal Distributions

Areas under a normal curve can be found using the NORMDIST function. The syntax is = NORMDIST$(x, \mu, \sigma,$ cumulative), where $\mu$ is the mean and $\sigma$ is the standard deviation. The parameter cumulative indicates whether the density (set cumulative = "false" or "0") or whether the cumulative distribution (set cumulative = "true" or "1") is wanted. The formula = NORMDIST$(x, \mu, \sigma, 1)$ returns $F(x)$, which is the area to the left of $x$ under an $N(\mu, \sigma)$ density and can be used to produce a table of normal areas as found in many statistics texts. Another formula, = NORMINV$(p, \mu, \sigma)$, returns the inverse $F^{-1}(p)$ of the cumulative, that is, a value $x$ such that the area to the left of $x$ is the specified $p$. For $N(0, 1)$, use NORMSDIST and NORMSINV instead.

# Macro – Normal Distribution Calculations

**Example 1.8.**    (Example 1.18, page 64 in the text.) The weights of 9-ounce potato chip bags are approximately Normal, with $\mu = 9.12$ ounces and $\sigma = 0.15$ ounces. What proportion of all 9-ounce bags weight less than 9.3 ounces?

**Solution**

Click on a cell (to activate it) where you want to locate the answer, say A1. Because we are looking for the area to the left of the point 9.3, the syntax is

$$\Phi(x) = \texttt{NORMDIST}(x, \mu, \sigma, 1).$$

Enter the formula = $\texttt{NORMDIST}(9.3, 9.12, 0.15, 1)$. The answer 0.8849 appears in cell A1. (Users of **Excel 5/95** may also use the **Function Wizard**, while users of **Excel 97/98** may use the **Formula Palette** to enter the formula.)

**Note:** We have included a simple macro *bCDF.xls* that does this. Complete the dialog box (shown to the left of Figure 1.34) and the answer appears in another box (to the left in this figure).

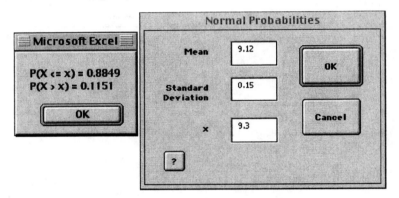

Figure 1.34: Normal Probabilities Macro

similarly upper tail areas can be calculated using

$$1 - \Phi(x) = 1 - \texttt{NORMDIST}(x, \mu, \sigma, 1)$$

while areas between points $a < b$ can be obtained from

$$\Phi(b) - \Phi(a) = \texttt{NORMDIST}(b, \mu, \sigma, 1) - \texttt{NORMDIST}(a, \mu, \sigma, 1)$$

**Example 1.9.**    (Example 1.22, "Backward" Normal calculation, page 69 in the text. ) Miles per gallon ratings of compact cars (2001 model year) follow approximately the $N(25.7, 5.88)$ distribution. How many miles per gallon must a vehicle get, to place in the top 10% of all 2001 model year compact cars?

**Solution.** We are looking for the mileage rating with area 0.1 to its right under the normal curve, equivalently with area 0.9 to its left. The Excel formula for the inverse of the cumulative normal distribution is

$$\Phi^{-1}(p) = \texttt{NORMINV}(p, \mu, \sigma).$$

Enter $= \texttt{NORMINV}(0.90, 25.7, 5.88)$ and read off 33.2 mpg, the 90th percentile.

## Graphing the Normal Curve

By combining the **ChartWizard** and the NORMDIST function, we can create a graph of any normal curve. *In fact, the procedure described here can be used to plot the graph of any function Excel can evaluate.*

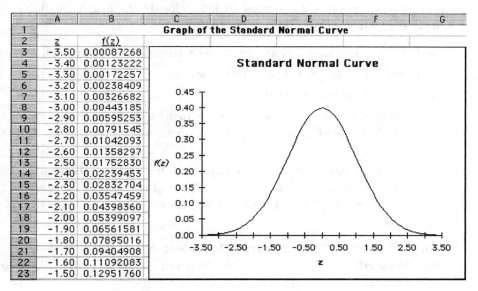

Figure 1.35: Graphing a Normal Density

**Example 1.10.** Construct a graph of the standard Normal curve.

**Solution**

1. Enter the labels $z$ and $f(z)$ in cells A2, B2. (See Figure 1.35.) $f(z)$ will represent the density or height of the normal curve at the point $z$.

2. Enter $-3.5$ and $-3.4$ in cells A3 and A4, respectively. These two points mark the ends of the range of values over which the normal density will be graphed. Next, we create a column of $z$ values at which the standard normal

density will be calculated. Select A3:A4, check the fill handle in the lower right corner of A4, and drag to cell A73 to fill the column with decreasing values of $z$ decremented by 0.1. Format the values with two decimal places.

3. Select cell B3 and enter $=$ NORMDIST(A3, 0, 1, 0) in the **Formula Bar**. Cell B3 now contains the value 0.00087268, the standard normal density evaluated at $z = -3.50$.

4. Select cell B3, click the fill handle, and drag down to B73. This copies the formula you just entered in B3 into cells B4:B73 relative to the corresponding cell references in column A. Column B is filled with values $f(z)$ of the standard normal density corresponding to each value of $z$ in column A.

5. **Users of Excel 5/95.** Select cells A2:B73, click the **ChartWizard** button, and then click in cell C2 and drag to I26 to locate the graph. A dialog box **ChartWizard – Step 1 of 5** appears.

   - In Step 1, you are given an opportunity to correct or confirm your range.
   - In Step 2, select **XY (Scatter)** chart.
   - In Step 3, select format **6**.
   - In Step 4, select the radio button **Columns** for Data Series, enter "1" for Column for Category (X) Axis Labels, and enter "1" for Row for Legend Text.
   - In Step 5, select the radio button **No** for **Add a Legend?**, type "Standard Normal Curve" as the **Chart Title**, and type $z$ and $f(z)$ for **Category (X)** and **Value (Y)** titles, respectively. Finally, click **Finish**.

   **Users of Excel 97/98/2000/2001.** Select cells A2:B73 and click the **ChartWizard**. A dialog box **ChartWizard – Step 1 of 4 – Chart Type** appears.

   - In Step 1, select **XY (Scatter)** for **Chart Type** and the lower right Chart sub-type **Scatter without markers**.
   - In Step 2, under the **Data Range** tab, the range will already be indicated and the **Series** radio button for **Columns** will be selected. You may edit the range if it is incorrect. Under the **Series** tab, no changes are necessary.
   - In Step 3, under the **Titles** tab, type "Standard Normal Curve" as the **Chart Title**, $z$ for **Value (X) Axis**, and $f(z)$ for **Value (Y) Axis**. Under the **Axes tab**, both check boxes should be selected. Under the **Gridlines** tab, clear all check boxes. Under the **Legend** tab, clear the **Show legend**. Finally, under the **Data Labels** tab, select the radio button **None**.

- In Step 4, embed the graph in the current workbook by selecting the radio button **As object in**. Finally, click **Finish**.

6. Activate the graph for editing, and format the display as you wish to present it, using the editing features discussed previously.

## Constructing a Normal Table

| z | 0.00 | 0.01 | 0.02 | 0.03 | 0.04 | 0.05 | 0.06 | 0.07 | 0.08 | 0.09 |
|------|--------|--------|--------|--------|--------|--------|--------|--------|--------|--------|
| 0.00 | 0.5000 | 0.5040 | 0.5080 | 0.5120 | 0.5160 | 0.5199 | 0.5239 | 0.5279 | 0.5319 | 0.5359 |
| 0.10 | 0.5398 | 0.5438 | 0.5478 | 0.5517 | 0.5557 | 0.5596 | 0.5636 | 0.5675 | 0.5714 | 0.5753 |
| 0.20 | 0.5793 | 0.5832 | 0.5871 | 0.5910 | 0.5948 | 0.5987 | 0.6026 | 0.6064 | 0.6103 | 0.6141 |
| 0.30 | 0.6179 | 0.6217 | 0.6255 | 0.6293 | 0.6331 | 0.6368 | 0.6406 | 0.6443 | 0.6480 | 0.6517 |
| 0.40 | 0.6554 | 0.6591 | 0.6628 | 0.6664 | 0.6700 | 0.6736 | 0.6772 | 0.6808 | 0.6844 | 0.6879 |
| 0.50 | 0.6915 | 0.6950 | 0.6985 | 0.7019 | 0.7054 | 0.7088 | 0.7123 | 0.7157 | 0.7190 | 0.7224 |
| 0.60 | 0.7257 | 0.7291 | 0.7324 | 0.7357 | 0.7389 | 0.7422 | 0.7454 | 0.7486 | 0.7517 | 0.7549 |
| 0.70 | 0.7580 | 0.7611 | 0.7642 | 0.7673 | 0.7704 | 0.7734 | 0.7764 | 0.7794 | 0.7823 | 0.7852 |
| 0.80 | 0.7881 | 0.7910 | 0.7939 | 0.7967 | 0.7995 | 0.8023 | 0.8051 | 0.8078 | 0.8106 | 0.8133 |
| 0.90 | 0.8159 | 0.8186 | 0.8212 | 0.8238 | 0.8264 | 0.8289 | 0.8315 | 0.8340 | 0.8365 | 0.8389 |
| 1.00 | 0.8413 | 0.8438 | 0.8461 | 0.8485 | 0.8508 | 0.8531 | 0.8554 | 0.8577 | 0.8599 | 0.8621 |
| 1.10 | 0.8643 | 0.8665 | 0.8686 | 0.8708 | 0.8729 | 0.8749 | 0.8770 | 0.8790 | 0.8810 | 0.8830 |
| 1.20 | 0.8849 | 0.8869 | 0.8888 | 0.8907 | 0.8925 | 0.8944 | 0.8962 | 0.8980 | 0.8997 | 0.9015 |
| 1.30 | 0.9032 | 0.9049 | 0.9066 | 0.9082 | 0.9099 | 0.9115 | 0.9131 | 0.9147 | 0.9162 | 0.9177 |
| 1.40 | 0.9192 | 0.9207 | 0.9222 | 0.9236 | 0.9251 | 0.9265 | 0.9279 | 0.9292 | 0.9306 | 0.9319 |
| 1.50 | 0.9332 | 0.9345 | 0.9357 | 0.9370 | 0.9382 | 0.9394 | 0.9406 | 0.9418 | 0.9429 | 0.9441 |
| 1.60 | 0.9452 | 0.9463 | 0.9474 | 0.9484 | 0.9495 | 0.9505 | 0.9515 | 0.9525 | 0.9535 | 0.9545 |
| 1.70 | 0.9554 | 0.9564 | 0.9573 | 0.9582 | 0.9591 | 0.9599 | 0.9608 | 0.9616 | 0.9625 | 0.9633 |
| 1.80 | 0.9641 | 0.9649 | 0.9656 | 0.9664 | 0.9671 | 0.9678 | 0.9686 | 0.9693 | 0.9699 | 0.9706 |
| 1.90 | 0.9713 | 0.9719 | 0.9726 | 0.9732 | 0.9738 | 0.9744 | 0.9750 | 0.9756 | 0.9761 | 0.9767 |
| 2.00 | 0.9772 | 0.9778 | 0.9783 | 0.9788 | 0.9793 | 0.9798 | 0.9803 | 0.9808 | 0.9812 | 0.9817 |
| 2.10 | 0.9821 | 0.9826 | 0.9830 | 0.9834 | 0.9838 | 0.9842 | 0.9846 | 0.9850 | 0.9854 | 0.9857 |
| 2.20 | 0.9861 | 0.9864 | 0.9868 | 0.9871 | 0.9875 | 0.9878 | 0.9881 | 0.9884 | 0.9887 | 0.9890 |
| 2.30 | 0.9893 | 0.9896 | 0.9898 | 0.9901 | 0.9904 | 0.9906 | 0.9909 | 0.9911 | 0.9913 | 0.9916 |
| 2.40 | 0.9918 | 0.9920 | 0.9922 | 0.9925 | 0.9927 | 0.9929 | 0.9931 | 0.9932 | 0.9934 | 0.9936 |
| 2.50 | 0.9938 | 0.9940 | 0.9941 | 0.9943 | 0.9945 | 0.9946 | 0.9948 | 0.9949 | 0.9951 | 0.9952 |
| 2.60 | 0.9953 | 0.9955 | 0.9956 | 0.9957 | 0.9959 | 0.9960 | 0.9961 | 0.9962 | 0.9963 | 0.9964 |
| 2.70 | 0.9965 | 0.9966 | 0.9967 | 0.9968 | 0.9969 | 0.9970 | 0.9971 | 0.9972 | 0.9973 | 0.9974 |
| 2.80 | 0.9974 | 0.9975 | 0.9976 | 0.9977 | 0.9977 | 0.9978 | 0.9979 | 0.9979 | 0.9980 | 0.9981 |
| 2.90 | 0.9981 | 0.9982 | 0.9982 | 0.9983 | 0.9984 | 0.9984 | 0.9985 | 0.9985 | 0.9986 | 0.9986 |
| 3.00 | 0.9987 | 0.9987 | 0.9987 | 0.9988 | 0.9988 | 0.9989 | 0.9989 | 0.9989 | 0.9990 | 0.9990 |
| 3.10 | 0.9990 | 0.9991 | 0.9991 | 0.9991 | 0.9992 | 0.9992 | 0.9992 | 0.9992 | 0.9993 | 0.9993 |
| 3.20 | 0.9993 | 0.9993 | 0.9994 | 0.9994 | 0.9994 | 0.9994 | 0.9995 | 0.9995 | 0.9995 | 0.9995 |
| 3.30 | 0.9995 | 0.9995 | 0.9995 | 0.9996 | 0.9996 | 0.9996 | 0.9996 | 0.9996 | 0.9996 | 0.9997 |
| 3.40 | 0.9997 | 0.9997 | 0.9997 | 0.9997 | 0.9997 | 0.9997 | 0.9997 | 0.9997 | 0.9997 | 0.9998 |
| 3.50 | 0.9998 | 0.9998 | 0.9998 | 0.9998 | 0.9998 | 0.9998 | 0.9998 | 0.9998 | 0.9998 | 0.9998 |
| 3.60 | 0.9998 | 0.9998 | 0.9999 | 0.9999 | 0.9999 | 0.9999 | 0.9999 | 0.9999 | 0.9999 | 0.9999 |
| 3.70 | 0.9999 | 0.9999 | 0.9999 | 0.9999 | 0.9999 | 0.9999 | 0.9999 | 0.9999 | 0.9999 | 0.9999 |
| 3.80 | 0.9999 | 0.9999 | 0.9999 | 0.9999 | 0.9999 | 0.9999 | 0.9999 | 0.9999 | 0.9999 | 0.9999 |
| 3.90 | 1.0000 | 1.0000 | 1.0000 | 1.0000 | 1.0000 | 1.0000 | 1.0000 | 1.0000 | 1.0000 | 1.0000 |

**Normal Table Constructed in Excel**

Figure 1.36: Normal Table

It is very easy in Excel to produce a table of normal areas. The method described below can be adapted to produce tables of other continuous distributions.

**Example 1.11.**   Construct a table of Normal areas (Figure 1.36).

**Solution**

1. Enter the label and values in column A and row 3. Column A is the first decimal while row 1 is the second decimal of $z$.

2. Enter the formula =NORMDIST($A4 + B$3) in cell B4. (Remember that the $ sign prefix makes the corresponding row or column label absolute.) Select cell B4, click the fill handle in the lower right corner of B4, and drag to K4.

3. Select cells B4:K4, click the fill handle in the lower right corner of K4, and drag to K43 to fill the block B4:K43.

## Macro – Normal Quantile Plots

There are several ways to assess whether a data set is normal. An analytic approach beyond the level of this book was developed by S. Shapiro and M. B. Wilk ("An analysis of variance test for normality", *Biometrika* **52**, pp. 591–611, 1965). A simple graphical approach constructs a histogram and compares the observed counts with the 68-95-99.7% rule. A more sensitive version of this idea is to order the observations and examine their distribution visually, using a scatterplot involving the corresponding expected quantiles of a normal curve. Normal data will tend to fall on a straight line. (This is the basis for the Shapiro-Wilk test.)

Because Excel does not provide a normal quantile plot, we have created a macro *PP.xls* for this purpose. It is based on the fact that the expected value of the $i$th order statistic (the $i$th largest in increasing magnitude) of a sample of size $n$ from a $N(0, 1)$ distribution can be approximated by the percentile

$$z_{(i)} = \text{NORMSINV}\left(\frac{i - \frac{3}{8}}{n + \frac{1}{4}}\right)$$

which is the value of a standard normal such that the area to the left is $\frac{i-\frac{3}{8}}{n+\frac{1}{4}}$. Our macro plots $z_{(i)}$ on the horizontal axis against $x_{(i)}$ on the vertical axis where $x_{(i)}$ is the $i$th largest from the data set $\{x_1, x_2, \ldots, x_n\}$.

**Example 1.12.** (Example 1.23, page 72 in text.) Construct a normal quantile plot of the earnings of the 15 black female hourly workers at National Bank (data in Figure 1.28).

**Solution**

1. Open the macro file *PP.xls* and the Excel file containing the data. From the Menu Bar select **Tools – Macro – Macros ...** to open the **Macro** dialog box, which lists the available macros. Highlight *PP.xls* to open the dialog box for the Normal Quantile Plot macro (Figure 1.37). Select the data range and click **OK**.

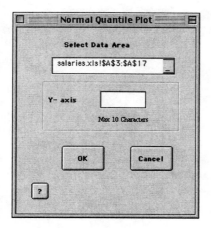

Figure 1.37: Normal Quantile Plot Macro Dialog Box

2. The quantile plot appears in another sheet of your workbook. It is an Excel chart and can be edited for appearance as in Figure 1.38

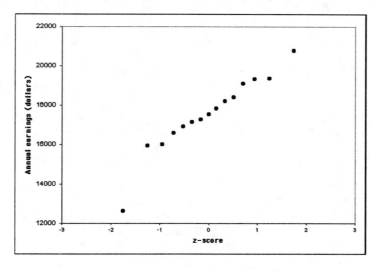

Figure 1.38: Normal Quantile Plot

# Chapter 2

# Examining Relationships

Response (dependent) variables measure the outcome of a study, while explanatory (independent) variables explain or predict changes in the response variable. It is important to examine graphically the nature of the relationship between these variables before imposing a specific mathematical model. Denote an explanatory variable by $x$ and a response variable by $y$. A plot in Cartesian coordinates of all pairs $(x_i, y_i)$ of observed values is called a **Scatterplot**.

## 2.1 Scatterplots

Figure 2.1 is a scatterplot constructed in Excel like the one on page 94 of the text. The scatterplot can be enhanced for display in a variety of ways by invoking the options available on the pull-down menu: for instance, colors of points, backround color, titles, labels of axes, labels of points, scale, presence or absence of gridlines. In fact, one of the nice features of Excel is its facility for achieving dramatic professional-looking graphs for reports or presentations. The steps involved in creating a scatterplot are similar to those for producing a **Histogram** using the **ChartWizard**. The instructions are based on **Excel 97/98/2000/2001**. **Excel 5/95** users should make corresponding changes.

> **Example 2.1.**　(Case 2.1, page 92 in text.) Three entrepreneurial women open Duck Worth Wearing, a shop selling high-quality second-hand children's clothing, toys, and furniture. Items are consigned to the shop by individuals who then receive a percentage of the selling price of their items. Table 2.1 below (also Table 2.1 on page 93 in text) displays daily data for April 2000 on sales of the gross sales and the number of items sold. Make a scatterplot of this data with number of items sold on the horizontal axis.

Table 2.1: Gross Sales and Number of Items Sold

| Date | Gross Sales ($) | Item Count |
|------|------|------|
| 4/1/00 | 890.5 | 115 |
| 4/3/00 | 197 | 17 |
| 4/4/00 | 231 | 26 |
| 4/5/00 | 170 | 21 |
| 4/6/00 | 202.5 | 30 |
| 4/7/00 | 225.5 | 35 |
| 4/8/00 | 489.7 | 84 |
| 4/10/00 | 234.8 | 42 |
| 4/11/00 | 161.5 | 21 |
| 4/12/00 | 284 | 44 |
| 4/13/00 | 422 | 65 |
| 4/14/00 | 300.7 | 59 |
| 4/15/00 | 412.4 | 69 |
| 4/17/00 | 346.8 | 59 |
| 4/18/00 | 92.3 | 19 |
| 4/19/00 | 255.8 | 42 |
| 4/20/00 | 118.5 | 16 |
| 4/21/00 | 286.5 | 39 |
| 4/22/00 | 594 | 72 |
| 4/24/00 | 263.29 | 43 |
| 4/25/00 | 244.08 | 45 |
| 4/26/00 | 394.28 | 64 |
| 4/27/00 | 241.31 | 36 |
| 4/28/00 | 299.97 | 40 |
| 4/29/00 | 649.04 | 103 |

## Creating a Scatterplot

Enter the data from Table 2.1 into consecutive columns of a worksheet (for instance in cells A1:B26).

Step 1. Select cells A1:B26 and click on the **ChartWizard**. From the choice of charts select **XY (Scatter)** for Chart type on the left and select the top Chart sub-type **Scatter** on the right. Click **Next**.

Step 2. The next dialog box previews the chart and allows any changes to be made to the data range (see Figure 2.1). Click **Next**.

Step 3. The **Chart Option** dialog box appears (Figure 2.2).

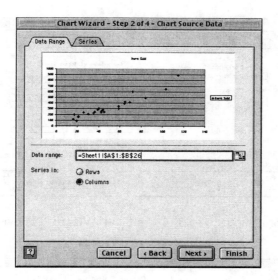

Figure 2.1: Step 2 of the ChartWizard

- Click the **Titles** tab and enter "Gross Sales vs. Items Sold" for **Chart title**, "Items Sold" for **Value (X) Axis**, and "Gross Sales" for **Value (Y) Axis**.

- Click the **Legend** tab. Clear the **Show Legend** check box. Click **Next**.

- Click the **Gridlines** tab and make sure that the **Major gridlines** box is checked for both axes. Click **Next**.

Step 4. In the last step select the radio button to enter the chart as an object on the current sheet. Click **Finish**. The scatterplot appears embedded in your workbook (Figure 2.3).

## Enhancing a Scatterplot

Every scatterplot may be enhanced using editing tools, some of which were described in Chapter 1. To see how this is done, activate the scatterplot by clicking once within its border. New commands become available that can be accessed under the Menu Bar. For instance, compare the pull-down options under **Insert**, **Format**, **Tools**, as well as the new **Chart**.

### Changing Scale

Excel uses a range from 0 to 100% as the default, and sometimes the scatterplot will show unwanted blank space. The scatterplot can be edited to change the

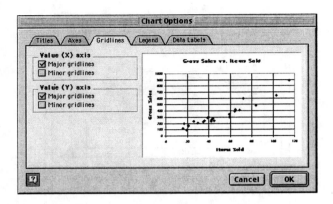

Figure 2.2: Step 3 of the ChartWizard – Preview

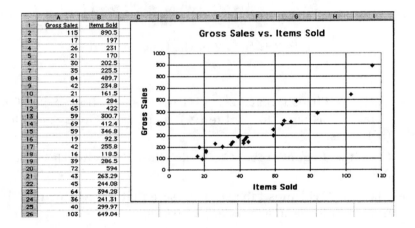

Figure 2.3: Data from Table 2.1 with Scatterplot

maximum or minimum X axis value by double clicking the X axis to produce the **Format Axis** dialog box. Equivalently select the X axis by clicking once and then choose **Format** from the Menu Bar. Make any changes you wish by selecting the appropriate tabs at the top of the dialog box. If desired, the Y axis may edited similarly. A scatterplot is a chart like the histogram and it may be enhanced in the same way that a histogram was enhanced in Section 1.1.

## Changing Titles

All titles on the scatterplot can be changed after it is created. Click and select the X axis title "Items Sold" and you can begin typing. If you double-click the X axis title, the **Format Axis Title** dialog box appears which allows you to change fonts, alignment, and "patterns" (which includes colors). The Y axis title and the chart title can be changed in the same manner.

## Labeling a Data Point

By default, Excel uses diamonds to plot the points. Suppose, for presentation purposes, you wish to use a different shape (and color) to represent a point and also to label a point. In particular, suppose you wish to label the point (72,594) with the corresponding date "04/22/2000". The following steps describe how to achieve this.

1. Activate the chart and drag the pointer to the point (72,594) and click. A text box will appear with the coordinates of the point (Figure 2.4).

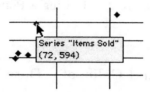

Figure 2.4: Locating a Point

2. Hold down the **Control** key (**Windows**) or **Command** key (**Macintosh**) and with your mouse pointer **select** the minimum point. Release the mouse button and select the point again. The pointer becomes a four-pointed "plus" sign (Figure 2.5). You can now access new commands under **Format** on the Menu Bar that let you edit the selected point.

Figure 2.5: Selecting a Point

3. For **Excel 5/95**, choose **Insert − Data Labels** from the Menu Bar to open the **Format Data Point** dialog box. For **Excel 97/98/2000/2001**, choose **Format − Selected Data Point...** from the Menu Bar to open a corresponding **Format Data Point** dialog box. Under the **Data Labels** tab select the radio button for **Show Value**. Click **OK**. Excel attaches the $y$ value "594" to this point on the scatterplot and encloses it within a bordered selection box ready for editing. Type "04/22/2000" (which appears

in the **Formula Bar**) and press **enter**. The selection box now contains the word "04/22/2000." You can select and move it, then **deselect** by clicking elsewhere. If you double-click on "04/22/2000" then the **Format Data Labels** dialog box appears, which provides additional editing features.

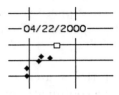

Figure 2.6: Editing a Point

## Changing the Marker and Color of a Data Point

Sometimes you may want to make a point stand out by changing its symbol and color on the scatterplot. We show how this is done using the "min" point above.

1. Activate the chart and click on the "min" observation as before.

2. Hold down the **Control** key (**Windows**) or **Command** key (**Macintosh**) and select the point so that the pointer again becomes a four-pointed plus sign.

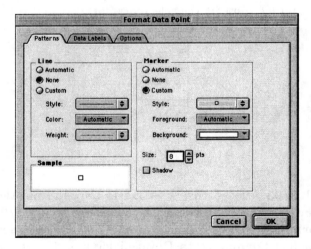

Figure 2.7: Changing the Default Marker

3. For **Excel 5/95**, choose **Insert – Data Labels** from the Menu Bar to open the **Format Data Point** dialog box. For **Excel 97/98/2000/2001**, choose **Format – Selected Data Point...** from the Menu Bar to open a corresponding **Format Data Point** dialog box. Under the **Patterns** tab, leave the **Line** selection as **None**. Under **Marker**, select a marker type from the pull-down list for **Style**, and also select a **Foreground** and **Background** color and size (Figure 2.7). In **Excel 5/95**, the size of the marker cannot be changed and there is no **Options** tab. Your selection is previewed in the small **Sample** box in the lower portion of the dialog box. Click **OK**. Figure 2.6 shows the result of the above editing.

## 2.2   Correlation

A scatterplot is a useful display of the nature of a relationship between two quantitative variables. It provides information about the functional form, the strength, and the direction. We next supplement this indicator of the relationship with a numerical measure.

The correlation between two variables $x$ and $y$ measures the strength of the linear association between them. For $n$ pairs of data points, $(x_i, y_i)$, $1 \leq i \leq n$, the sample correlation coefficient is defined to be

$$r = \frac{1}{n-1} \sum_{i=1}^{n} \left( \frac{x_i - \bar{x}}{s_x} \right) \left( \frac{y_i - \bar{y}}{s_y} \right)$$

where $\bar{x}$ and $\bar{y}$ are the sample means of the $\{x_i\}$ and $\{y_i\}$, respectively, and $s_x$ and $s_y$ are the corresponding sample standard deviations. Excel provides several ways to calcualate the correlation coefficient.

### The CORREL Function

The most direct way to find the correlation for Example 2.1 is by the **Formula Palette** for **Excel 97/98/2000/2001**, or the **Function Wizard** for **Excel 5/95** with the function CORREL, which computes the correlation coefficient.

> **Example 2.3.**   (Exercise 2.20, page 110 in text.) Find the correlation $r$ between the size in gigabytes and the price in dollars of several external hard-drive models from one manufacturer.

Table 2.2: The Price of a Gigabyte

| Size (GB) | 8 | 6 | 18 | 30 | 20 | 10 | 3 |
|-----------|-----|-----|-----|-----|-----|-----|-----|
| Price ($) | 310 | 290 | 500 | 800 | 470 | 330 | 150 |

**Solution**

1. Enter the above data into two adjacent columns in a worksheet, say, cells A3:B9. Select an empty cell where you want the correlation to appear and invoke either the **Function Wizard** or the **Formula Palette**. In each case, select **Statistical** for Function Category and CORREL for Function Name.

2. Enter A3:A9 for **Array1** and B1:B6 for **Array2**. You may enter by hand or click and drag on the workbook over the range A3:A9, press **Tab** on the keyboard, then click and drag over the range B3:B9, and finally click **OK** (or **Finish**, for **Excel 5/95**). The answer 0.9793 appears in the cell you selected.

## The Correlation Tool

Correlation between two variables can also be calculated using the **Correlation** tool in the **Analysis ToolPak**. This tool is most effective, however, for determining pairwise correlations for multivariate data sets, for which repeated use of the CORREL function would be inefficient. This tool prints out a matrix of correlations. Such a matrix is helpful in multiple regression to decide which variables to include in a model.

Table 2.3: The 10 Largest Online Brokerages

| Brokerage | Market Share | Accounts | Assets |
|---|---|---|---|
| Charles Schwab | 27.5 | 2500 | 219 |
| E* Trade | 12.9 | 909 | 21.1 |
| TD Waterhouse | 11.6 | 615 | 38.8 |
| Datek | 10 | 205 | 5.5 |
| Fidelity | 9.3 | 2300 | 160 |
| Ameritrade | 8.4 | 428 | 19.5 |
| DLJ Direct | 3.6 | 590 | 11.2 |
| Discover | 2.8 | 134 | 5.9 |
| Suretrade | 2.2 | 130 | 1.3 |
| National Discount Brokers | 1.3 | 125 | 6.8 |

**Example 2.4.** (Exercise 11.1, page 648 in text.) Because this tool is most useful for multivariate data, we have chosen an example from multiple regression in Chapter 11 in the text to illustrate how to use it. Table 2.3 (which is Table 11.3 on page 649 in the text) gives data on market share, number of accounts, and assets held by the 10 largest online stock brokerages. Market share is expressed in percent, based

on the number of trades per day. The number of accounts is given in thousands, and the assets are in billions of dollars. Find the correlations between the variables "Market Share," "Accounts," and "Assets."

**Solution.** Enter the data into four columns of a worksheet, say A1:D11.

| | A | B | C | D | E | F | G | H |
|---|---|---|---|---|---|---|---|---|
| 1 | Brokerage | Market Share | Accounts | Assets | | *Market Share* | *Accounts* | *Assets* |
| 2 | Charles Schwab | 27.5 | 2500 | 219 | Market Share | 1 | | |
| 3 | E* Trade | 12.9 | 909 | 21.1 | Accounts | 0.75318 | 1 | |
| 4 | TD Waterhouse | 11.6 | 615 | 38.8 | Assets | 0.78017 | 0.96827 | 1 |
| 5 | Datek | 10 | 205 | 5.5 | | | | |
| 6 | Fidelity | 9.3 | 2300 | 160 | | | | |
| 7 | Ameritrade | 8.4 | 428 | 19.5 | | | | |
| 8 | DLJ Direct | 3.6 | 590 | 11.2 | | | | |
| 9 | Discover | 2.8 | 134 | 5.9 | | | | |
| 10 | Suretrade | 2.2 | 130 | 1.3 | | | | |
| 11 | National Discount Brokers | 1.3 | 125 | 6.8 | | | | |

Figure 2.8: Correlation ToolPak Output

1. From the Menu Bar choose **Tools − Data Analysis**, and in the dialog box highlight **Correlation** and click **OK**. A dialog box **Correlation** appears.

2. In this dialog box, enter B1:D11 for **Input range** (most conveniently done by clicking and dragging over this range on the workbook and pressing the Tab key). Remember not to include data from Column A because there are no correlations involving names of the Brokerages. Check the box **Labels in first row**, point to cell E1 for **Output range**, and click **OK**.

### Excel Output

The output appears in E1:H4, as shown in Figure 2.8. In view of symmetry, only half the correlation matrix is required. We can read off the correlations: 0.753 between Accounts and Market Share; 0.780 between Assets and Market Share; and 0.968 between Assets and Accounts.

## Macro − Correlation Simulation

The macro *corr.xls* simulates data that comes from a bivariate normal distribution with means 0, standard deviations 0, and a general correlation coefficient $\rho$, which the student may change by the scroll bar. The current value of the population correlation coefficient appears in cell E38. The sample correlation coefficient appears in cell D42. The **Resample** button allows the user to simulate repeatedly and develop a feeling for the degree of linear association suggested by any particular value of a correlation coefficient. In particular, plots akin to those shown in Figure 2.10 on page 112 of the text may easily be obtained merely by changing the value of $\rho$. The macro is protected so it will not be inadvertently changed when used. Both

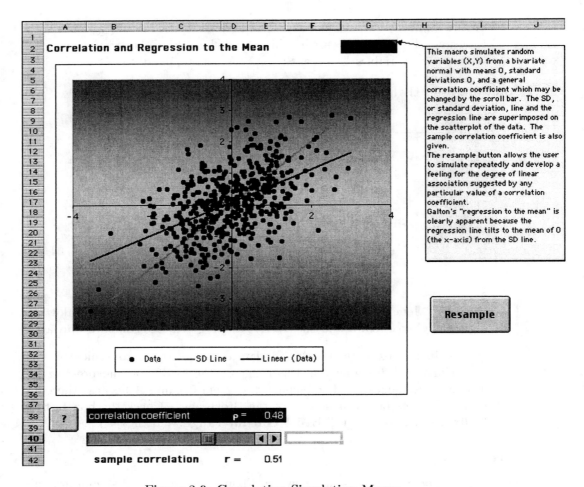

Figure 2.9: Correlation Simulation Macro

the least-squares regression line and the SD (standard deviation) line (which joins points at the same percentile or standard deviation, since the standard deviations are equal) are shown. Galton's "regression to the mean" is clearly apparent because the regression line tilts towards the mean of 0 (the X-axis) from the SD line.

## 2.3 Least-Squares Regression

We have seen how to plot two variables against each other in a scatterplot and have calculated the correlation coefficient to measure the strength of the linear association between them. It is useful to have an analytic relationship between the explanatory variable $x$ and the response variable $y$ of the form

$$y = f(x)$$

for predicting $y$ from $x$. Such a relationship is called a simple (meaning one explanatory variable) **regression curve**. The simplest curve is a straight line

$$y = a + bx$$

called the regression line of $y$ on $x$. The regression line represents, under certain assumptions, the mean response at each specified value $x$.

The method used to determine the coefficients $a$ and $b$ goes back at least to the great mathematician Gauss and is called the **Principle of Least-Squares**. Gauss himself recognized that the criterion was arbitrary and he used it because the coefficients $a$ and $b$ were then solvable in closed form. (Additional reasons connected with the errors being normal are presented in more advanced treatments.)

For a given $x_i$, we call

$$\hat{y}_i = a + bx_i$$

the **predicted** value and

$$e_i = y_i - \hat{y}_i$$

the **residual**. The **error sum of squares** is defined to be

$$\sum_{i=1}^{n} e_i^2 = \sum_{i=1}^{n} (y_i - a - bx_i)^2$$

By differentiating with respect to $a$ and $b$, we can solve for the values that minimize $\sum_{i=1}^{n} e_i^2$. These are the values used in the regression line. They are given by the formulas

$$\begin{aligned} \text{slope} \quad b &= r\,\frac{s_y}{s_x} \\ \text{intercept} \quad a &= \bar{y} - b\bar{x} \end{aligned}$$

where $\bar{x} = \frac{1}{n}\sum_{i=1}^{n} x_i$, $\bar{y} = \frac{1}{n}\sum_{i=1}^{n} y_i$, $r$ is the correlation coefficient, and

$$\begin{aligned} (n-1)s_x^2 &= \sum_{i=1}^{n}(x_i - \bar{x})^2 = \sum_{i=1}^{n} x_i^2 - \frac{1}{n}\left(\sum_{i=1}^{n} x_i\right)^2 \\ (n-1)s_y^2 &= \sum_{i=1}^{n}(y_i - \bar{y})^2 = \sum_{i=1}^{n} y_i^2 - \frac{1}{n}\left(\sum_{i=1}^{n} y_i\right)^2 \end{aligned}$$

## Fitting a Line to Data

Excel provides three built-in methods for regression analysis: **Trendline**, the **Regression** tool in the **Analysis ToolPak**, and regression functions such as FORECAST and TREND. For merely graphing a regression line and providing its equation and the coefficient of determination $r^2$, the **Trendline** command suffices. We will consider the **Regression** tool in Chapters 10 and 11, as well as regression functions.

### Linear Trendline

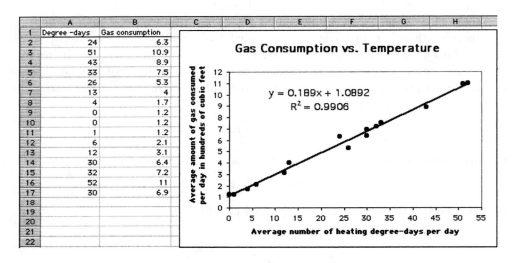

Figure 2.10: Gas Consumption Scatterplot with Linear Trendline

We use the **Linear Trendline** to insert a curve on a scatterplot. The trendline can be added to any scatterplot even after the **Regression** tool is used.

> **Example 2.5.** (Example 2.4 page 96 and Examples 2.8–2.9, pages 117–121 in text.) The Sanchez household is about to install solar panels to reduce the cost of heating their house. To know how much the solar panels help, they record consumption of natural-gas before the panels are installed. Gas consumption is higher in cold weather, so the relationship between outside temperature and gas consumption is important. Table 2.2 on page 96 in the text give data for 16 months. The response variable $y$ is the average amount of natural-gas consumed each day during the month, in hundreds of cubic feet. The explanatory variable $x$ is the average number of heating degree-days each day during the month. Fit the least-squares regression line to the data.

**Solution.**    We first construct a scatterplot of the data (also shown in Figure 2.10) to verify that a linear model is appropriate.

1. Enter the data in cells A1:B17 in a worksheet. Use the **ChartWizard** to create a scatterplot and edit it for appearances so that it approximates what is shown in Figure 2.10. The scatterplot shows an approximate linear relationship, so it is appropriate to fit the data pairs with a straight line.

2. Activate the chart for editing and select the data by **clicking on one of the points**. The points appear highlighted, the **Name** box in the **Formula Bar** shows Series "Gas consumption," and in the text entry area we can read

   $$= \texttt{SERIES}(\text{Sheet1!}\$B\$1, \text{Sheet1!}\$A\$2 : \$A\$17, \text{Sheet1!}\$B\$2 : \$B\$17, 1)$$

   meaning that the series has been selected. (Refer to the online help for more information on this function and the Introduction for the meaning of Sheet1! notation.)

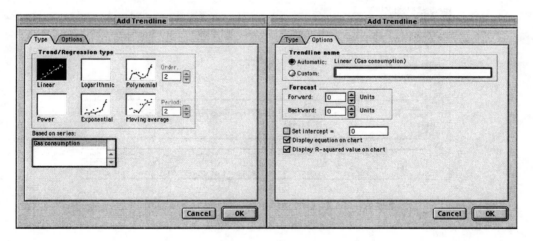

Figure 2.11: Trendline Type

3. For **Excel 5/95**, choose **Insert − Trendline** from the Menu Bar; for **Excel 97/98/2000/2001**, choose **Chart − Add Trendline...** from the Menu Bar. Then proceed as follows. Click the **Type** tab and select **Linear** (Figure 2.11). **Excel 97/98/2000/2001** have an additional text area (**Based on series**) in the blank space at the bottom of this figure. Click the **Options** tab and select the radio button **Automatic:Linear (Gas consumption)**. Check the boxes **Display equation on chart** and **Display R-squared value on dhart**. Make sure that the **Set intercept** box is clear. Click OK. The regression line is superimposed on the scatterplot, its equation $y =$

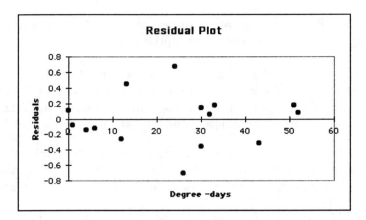

Figure 2.12: Residual Plot

| | J | K | L | M | N | O | P |
|---|---|---|---|---|---|---|---|
| 1 | SUMMARY OUTPUT | | | | | | |
| 2 | | | | | | | |
| 3 | *Regression Statistics* | | | | | | |
| 4 | Multiple R | 0.99526401 | | | | | |
| 5 | R Square | 0.99055044 | | | | | |
| 6 | Adjusted R Sq | 0.98987547 | | | | | |
| 7 | Standard Erro | 0.3389284 | | | | | |
| 8 | Observations | 16 | | | | | |
| 9 | | | | | | | |
| 10 | ANOVA | | | | | | |
| 11 | | *df* | *SS* | *MS* | *F* | *Significance F* | |
| 12 | Regression | 1 | 168.581161 | 168.581161 | 1467.55072 | 1.4152E-15 | |
| 13 | Residual | 14 | 1.60821443 | 0.11487246 | | | |
| 14 | Total | 15 | 170.189375 | | | | |
| 15 | | | | | | | |
| 16 | | *Coefficients* | *Standard Error* | *t Stat* | *P-value* | *Lower 95%* | *Upper 95%* |
| 17 | Intercept | 1.08921084 | 0.13891466 | 7.84086311 | 1.729E-06 | 0.79126826 | 1.38715342 |
| 18 | Degree -days | 0.18899895 | 0.00493359 | 38.3086246 | 1.4152E-15 | 0.17841745 | 0.19958046 |

Figure 2.13: Output from Regression Tool

$0.189x + 1.0892$ is displayed, and the coefficient of determination $R^2 = 0.9906$ (in Excel's notation) is inserted on the scatterplot.

4. The output may be edited as previously with other charts for presentation purposes. For instance, activate the chart and click on the rectangular box surrounding the equation; the border turns a darker grey. Use the **Decimal** tool to increase or decrease the number of decimal points. You can edit the text by replacing $x$ with "degree-days" and $y$ with "gas consumption." Move the equation and coefficient of determination to any convenient place. The final result with the regression line was previously shown in Figure 2.10.

# Residuals

No discussion of regression is complete without an analysis of residuals, which provide evidence of how well the regression model fits. Figure 2.12 shows scatterplot of residuals against degree-days for Example 2.5. There does not appear to be any discernable pattern in the plot, indicating that a straight-line fit is appropriate. We defer discussion of this topic to Chapter 10 where the **Regression** tool will be introduced but we will show in Figure 2.13 the output of the Regression. This may be compared with Figure 2.13 on page 121 in the text.

# Chapter 3

# Producing Data

Data represent populations. Whether obtained by an observational study or by a well-designed experiment, data represent the raw material for drawing sound conclusions about populations. It is implicit that the data reflect the population adequately, which means that the data must be a representative **sample** from the population.

The basic requirement of a sample is that it be a probability sample, meaning that the manner in which it was obtained can be quantified using probability (discussed in the next chapter). Samples can be obtained in a variety of ways most of which are built upon the simple random sample.

**Definition.** *A simple random sample (SRS) of size n consists of n individuals from the population chosen in such a way that every subset of size n individuals has an equal chance to be in the sample actually selected.* In Excel the RAND() function can be used to obtain an SRS. This of course requires that the population values be entered into a worksheet.

Excel also provides in its Data Analysis Toolpak, the **Sampling** and the **Random Number Generation** tools for generating samples from a specified population. These give sufficient flexibility for sampling from nearly any population. Repeated use of sampling leads to **simulation**, a useful technique for deriving properties of statistics and estimates. This is discussed in detail in Chapter 4. In this chaper we consider both sampling without replacement and sampling with replacement (SRS) from a specified population.

## 3.1   Simple Random Samples (SRS)

### The RAND() Function

To obtain an SRS, Excel uses the function RAND() to repeatedly select random uniform (0,1) numbers and assign them to members of a population. These numbers are then sorted to obtain a random permutation of the population from which an SRS of any desired size. This is tantamount to physically reaching into the population and scooping out a random sample.

**Example 3.1.** (Examples 3.5 and 3.7, pages 184–186 in text.) Joan's accounting firm serves 30 small business clients. Joan wants to interview a sample of five clients in detail to find ways to improve client satisfaction. To avoid bias, she chooses an SRS of size 5. Use Excel to find a simple random sample.

**Solution.** The text shows how to do this using Table B of random digits, the first step of which requires giving each client a numberical label with as few digits as possible. Excel also begins this way. The list of clients and their labels is in Table 3.1.

Table 3.1: Joan's Clients and Labels

| 01 | A-1 Plumbing | 16 | JL Records |
|----|----|----|----|
| 02 | Accent Printing | 17 | Johnson Commodities |
| 03 | Action Sport Shop | 18 | Keiser Construction |
| 04 | Anderson Construction | 19 | Liu's Chinese Restaurant |
| 05 | Bailey Trucking | 20 | Magic Tan |
| 06 | Balloons Inc. | 21 | Peerless Machine |
| 07 | Bennett Hardware | 22 | Photo Arts |
| 08 | Best's Camera Shop | 23 | River City Books |
| 09 | Blue Print Specialties | 24 | Riverside Tavern |
| 10 | Central Tree Service | 25 | Rustic Boutique |
| 11 | Classic Flowers | 26 | Satellite Services |
| 12 | Computer Answers | 27 | Scotch Wash |
| 13 | Darlene's Dolls | 28 | Sewing Center |
| 14 | Feisch Realty | 29 | Tire Specialties |
| 15 | Hernandez Electronics | 30 | Von's Video Store |

1. Give each client a unique numerical label from the set $\{1, 2, 3, \ldots, 30\}$ and enter the values into cells A4:A33 of a workbook (Figure 3.1). Enter the label "Clients" in cell A3 and "RAND()" in cell B3.

2. Enter "= RAND()" in cell B4 and fill down to B33. The function RAND() selects a number uniformly in (0,1).

3. Select cells B4:B33 and from the Menu Bar choose **Edit – Copy**. Then, with B4:B33 still selected, choose **Edit – Paste Special** from the Menu Bar. (**Windows** users can click the **right mouse button** while **Macintosh** users should hold down the **Option – Command** keys or the **Control** key

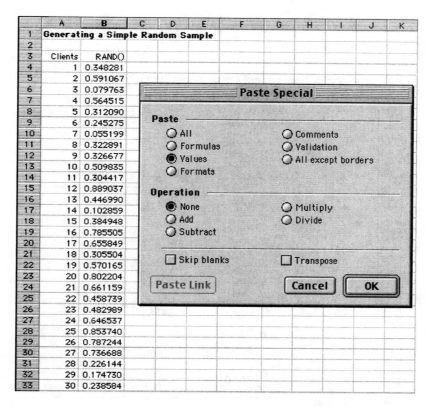

Figure 3.1: Original Labels and Paste Special Dialog Box

and click to get the **Shortcut Menu** box.) Select **Paste − Special**, and in the dialog box (also shown in Figure 3.1) select the radio buttons for **Values** and **None**. This replaces the formulas in the cells of column B by the actual values they take. This step is needed because, otherwise, actions taken in the workbook will recalculate the RAND() values.

4. Select cells A3:B34, from the Menu Bar choose **Data − Sort**, and in the **Sort By** drop-down list, click the arrow and select **Sample**. Also, select the radio button for **Header Row** (Figure 3.2). Excel sorts the data in ascending order in column B and carries the order to column A, which gives a random permutation of column A.

5. The sorted data appear in Figure 3.2. Designate the first five cells, A4:A8, to determine the random sample of five clients. Of course, any five can be selected once the sort has occurred.

6. The sample is the group of clients labelled 7, 3, 14, 29, 28. These are Bennett Hardware, Action Sport Shop, Feisch Realty, Sewing Center, and

Tire Specialties. Your own selection will be different because you are taking a random sample.

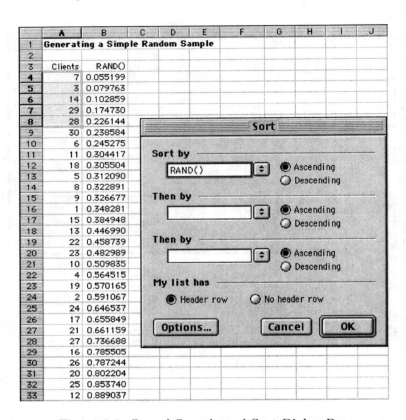

Figure 3.2: Sorted Sample and Sort Dialog Box

## 3.2   Samples with Replacement

### Using the Sampling Tool

Sometimes a random sample with replacement is needed. This may be viewed as an SRS from an infinite population. In Excel, such a sample is obtained by using the **Sampling Tool**. The user defines the population from which the sample is taken. For some of the standard populations (defined by the distributions of random variables in Chapter 4) there is a better method using the **Random Number Generation** tool.

**Example 3.2.**   (Based loosely on the NBA Draft Lottery from 1990 to 1993). In October of 1989, the Board of Governors of the NBA adopted a weighted system to begin with the 1990 NBA Draft Lottery. The team with the worst record during the regular season received 11 chances at the top pick (out of a total of 66), the second-worst team got 10 chances, and so on until the team with the best record among the non-playoff clubs got one chance. The order of picks was determined by the order in which the ping pong balls were selected. This lottery determined the order of the first three teams only. The remaining non-playoff teams then selected in inverse order of their regular season records. The lottery was an attempt to equalize the strength in teams, because under this system, the team with the worst record was assured of picking no worse than fourth. Although the lottery is clearly sampling without replacement, we will use the population of $1 + 2 + \ldots + 11 = 66$ ping pong balls to select a sample of size 3 with replacement.

**Solution.**

1. Enter the digits $\{1, 2, 3, 4, 5, 6, 7, 8, 9, 10, 11\}$ in A4:F14 of a worksheet with multiplicities as described by the lottery rules (Figure 3.3).

| | A | B | C | D | E | F | G | H |
|---|---|---|---|---|---|---|---|---|
| 1 | Sampling With Replacement | | | | | | | |
| 2 | | | | | | | | |
| 3 | | | | Population | | | | Teams Selected |
| 4 | 1 | 5 | 7 | 8 | 9 | 11 | | 4 |
| 5 | 2 | 5 | 7 | 8 | 10 | 11 | | 8 |
| 6 | 2 | 5 | 7 | 8 | 10 | 11 | | 9 |
| 7 | 3 | 5 | 7 | 9 | 10 | 11 | | |
| 8 | 3 | 6 | 7 | 9 | 10 | 11 | | |
| 9 | 3 | 6 | 7 | 9 | 10 | 11 | | |
| 10 | 4 | 6 | 8 | 9 | 10 | 11 | | |
| 11 | 4 | 6 | 8 | 9 | 10 | 11 | | |
| 12 | 4 | 6 | 8 | 9 | 10 | 11 | | |
| 13 | 4 | 6 | 8 | 9 | 10 | 11 | | |
| 14 | 5 | 7 | 8 | 9 | 10 | 11 | | |

Figure 3.3: Sample with Replacement

2. From the Menu Bar choose **Tools − Data Analysis** and select **Sampling** from the dialog box. Click **OK**.

3. Complete the **Sampling** dialog box as shown in Figure 3.4 and click **OK**. A random sample of size 4 appears in cells H4:H7.

# Random Digits

Figure 3.5 is a table of random digits, a list of the digits $\{0, 1, \ldots, 9\}$ that has the following properties:

1. The digits in all positions in the list have the same chance of being any one of $\{0, 1, \ldots, 9\}$.

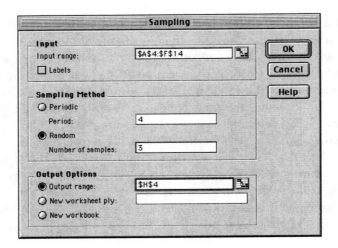

Figure 3.4: Sampling Dialog Box

2. The digits in different positions are independent in the sense that the value of one has no influence on the value of any other.

You can imagine asking an assistant (or computer) to mix the digits $\{0, 1, \ldots, 9\}$ in a hat, draw one, then replace the digit drawn, mix again, draw a second digit, and so on. We did something like this in the previous section. In **Excel 97/98**, this table of random digits dynamically changes when the F9 key on the keyboard is pressed. In **Excel 2000/2001**, you need to hold down the **Option** key (**Mac**) or **Control** key (**Windows**) at the same time.

### Using the RAND() Function

The Excel function INT truncates a real number to its integer value. For instance, the formula = INT(3.82) produces the value 3. By combining INT and RAND() as = INT(10*RAND()) we can produce random digits from $\{0, 1, \ldots, 9\}$.

**Example 3.3.** Produce a table of 500 random digits.

**Solution**

1. In cell A4 of a workbook, enter the formula = INT(10*RAND()).

2. Select cell A4, click the fill handle in the lower right corner, and drag across to cell T4.

3. Select cells A4:T4, click the fill handle in the lower right corner, and drag down to cell T28.

The result is shown in Figure 3.5.

| | A | B | C | D | E | F | G | H | I | J | K | L | M | N | O | P | Q | R | S | T |
|---|---|---|---|---|---|---|---|---|---|---|---|---|---|---|---|---|---|---|---|---|
| 1 | | | | | | | | Table of Random Digits | | | | | | | | | | | | |
| 2 | | | | | | | | =INT(10*(RAND())) | | | | | | | | | | | | |
| 3 | | | | | | | | | | | | | | | | | | | | |
| 4 | 6 | 1 | 7 | 5 | 6 | 7 | 5 | 8 | 7 | 4 | 1 | 2 | 4 | 7 | 3 | 2 | 9 | 5 | 2 | 1 |
| 5 | 7 | 4 | 9 | 7 | 5 | 2 | 5 | 0 | 8 | 7 | 7 | 0 | 5 | 7 | 2 | 2 | 2 | 6 | 5 | 8 |
| 6 | 0 | 9 | 3 | 2 | 4 | 2 | 2 | 8 | 2 | 3 | 0 | 5 | 4 | 9 | 1 | 3 | 5 | 3 | 6 | 4 |
| 7 | 6 | 2 | 3 | 2 | 9 | 0 | 6 | 2 | 3 | 1 | 4 | 4 | 7 | 0 | 8 | 9 | 8 | 8 | 9 | 2 |
| 8 | 6 | 9 | 4 | 5 | 7 | 7 | 9 | 7 | 3 | 4 | 5 | 1 | 7 | 8 | 8 | 4 | 2 | 1 | 0 | 0 |
| 9 | 6 | 4 | 8 | 7 | 4 | 5 | 8 | 0 | 2 | 0 | 5 | 8 | 5 | 8 | 8 | 9 | 6 | 6 | 8 | 7 |
| 10 | 1 | 4 | 9 | 7 | 7 | 0 | 2 | 6 | 0 | 0 | 4 | 2 | 2 | 9 | 4 | 3 | 0 | 6 | 0 | 5 |
| 11 | 6 | 8 | 9 | 0 | 3 | 9 | 3 | 0 | 0 | 7 | 6 | 9 | 2 | 1 | 7 | 5 | 4 | 8 | 3 | 3 |
| 12 | 3 | 7 | 1 | 4 | 5 | 4 | 7 | 2 | 5 | 4 | 5 | 4 | 9 | 3 | 2 | 3 | 9 | 9 | 9 | 8 |
| 13 | 7 | 7 | 9 | 3 | 4 | 6 | 6 | 8 | 6 | 7 | 1 | 5 | 0 | 9 | 3 | 4 | 0 | 3 | 0 | 4 |
| 14 | 3 | 2 | 8 | 7 | 0 | 4 | 4 | 9 | 3 | 1 | 6 | 5 | 0 | 7 | 7 | 1 | 1 | 5 | 9 | 1 |
| 15 | 3 | 1 | 5 | 7 | 0 | 0 | 5 | 2 | 0 | 3 | 5 | 3 | 3 | 6 | 6 | 3 | 8 | 6 | 1 | 8 |
| 16 | 2 | 1 | 1 | 2 | 4 | 4 | 8 | 1 | 3 | 2 | 5 | 2 | 2 | 3 | 1 | 7 | 7 | 4 | 5 | 1 |
| 17 | 6 | 6 | 2 | 6 | 8 | 2 | 9 | 6 | 6 | 4 | 9 | 5 | 7 | 2 | 7 | 8 | 5 | 9 | 3 | 6 |
| 18 | 2 | 3 | 7 | 0 | 2 | 7 | 5 | 7 | 0 | 2 | 5 | 6 | 5 | 6 | 3 | 1 | 1 | 9 | 2 | 1 |
| 19 | 4 | 1 | 7 | 7 | 9 | 7 | 5 | 8 | 1 | 3 | 8 | 5 | 3 | 1 | 1 | 8 | 9 | 1 | 9 | 1 |
| 20 | 1 | 3 | 2 | 4 | 3 | 6 | 1 | 3 | 1 | 5 | 0 | 7 | 2 | 3 | 7 | 4 | 7 | 5 | 9 | 1 |
| 21 | 6 | 8 | 4 | 4 | 2 | 6 | 8 | 7 | 0 | 7 | 5 | 2 | 5 | 0 | 2 | 7 | 5 | 4 | 5 | 1 |
| 22 | 0 | 7 | 3 | 6 | 7 | 8 | 4 | 5 | 4 | 1 | 8 | 1 | 5 | 7 | 6 | 5 | 3 | 9 | 6 | 2 |
| 23 | 7 | 7 | 3 | 3 | 3 | 2 | 5 | 9 | 9 | 6 | 3 | 6 | 3 | 1 | 0 | 8 | 0 | 0 | 7 | 2 |
| 24 | 8 | 9 | 3 | 1 | 3 | 6 | 8 | 3 | 4 | 4 | 7 | 4 | 8 | 8 | 0 | 7 | 0 | 5 | 0 | 3 |
| 25 | 2 | 3 | 2 | 2 | 6 | 7 | 4 | 6 | 4 | 1 | 2 | 7 | 9 | 6 | 2 | 9 | 3 | 1 | 9 | 1 |
| 26 | 3 | 0 | 3 | 8 | 4 | 2 | 4 | 2 | 0 | 3 | 0 | 0 | 5 | 4 | 5 | 4 | 7 | 8 | 9 | 6 |
| 27 | 1 | 5 | 2 | 9 | 3 | 6 | 1 | 4 | 8 | 8 | 2 | 2 | 8 | 1 | 1 | 0 | 0 | 3 | 1 | 1 |
| 28 | 6 | 5 | 0 | 1 | 0 | 0 | 3 | 4 | 4 | 9 | 8 | 9 | 9 | 0 | 1 | 0 | 9 | 6 | 1 | 6 |

Figure 3.5: Table of Random Digits

## 3.3 Toward Statistical Inference

A **parameter** is a number that describes the population. A **statistic** is a number that describes a sample. The value of a statistic changes from sample to sample while the parameter stays constant. An **estimate** is a statistic that is chosen as a guess for the value of a parameter.

The choice of a statistic for estimation is dependent on its sampling distribution, which describes the ensemble of values the statistic takes in repeated samples.

**Example 3.3.** (Case 3.1, page 216 and Example 3.20, page 218 in text.) Changing consumer attitudes toward shopping are of great interest to retailers and makers of consumer goods. The Yankelovich market research firm specializes in the study of "consumer trends." The firm conducts a nationwide survey, taking a random sample of 2,500 adults and asking if they agree or disagree that shopping is often frustrating and time-consuming. Of the respondents, 1,650 said they agreed. The proportion of the Yankelovich sample who agreed that clothes shopping is often frustrating is

$$\hat{p} = \frac{1650}{2500} = 0.66 = 66\%$$

The number $\hat{p} = 0.66$ is a *statistic*. The corresponding *parameter* is the proportion $p$ of all adult U.S. residents who would have agreed if asked the same question. We use $\hat{p}$ as an estimate of the unknown $p$. If the Yankelovich survey took a second random sample of 2,500 adults, the new sample would have different people and the corresponding value of $\hat{p}$ would be different. Suppose that in fact (unknown to Yankelovich) exactly 60% of all adults find clothes shopping frustrating and time-consuming. Suppose we select an SRS of size $n = 100$ and calculate $\hat{p}$. How variable are the values of $\hat{p}$ in repeated samples?

## Macro – Sampling Distribution

We answer this question with a macro *samdis.xls*, which constructs a histogram of 1000 simulated values of $\hat{p}$. The value of $p$ is initially set at 0.60. Using the **Spinner** control, this can be changed within the range $[0.50, 0.70]$. The value of $\hat{p}$ is also given, as well as the error $\hat{p} - p$. The histogram shows the range of values taken by $\hat{p}$ as well as its *sampling* distribution. Furthermore, if the **Resample** button is clicked, then a new histogram appears with another 1000 simulated values. By repeated resampling, one observes that all histograms are generally symmetric and bell shaped with a center near the true value $p$ (Figure 3.6).

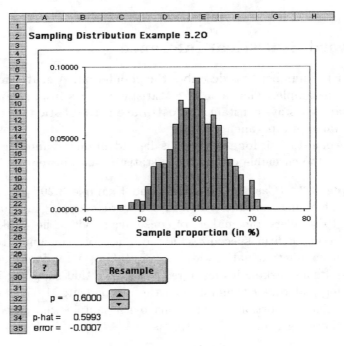

Figure 3.6: Sampling Distribution

# Chapter 4

# Probability and Sampling Distributions

A phenomenon is called random if individual outcomes are uncertain but there is still a regular distribution of outcomes in a large number of repetitions. The probability of any outcome of a random phenomenon is the proportion of times the outcome would occur in a very long series of repetitions. A real-world probability can only be estimated through the observation of data.

One way to develop an intuition for randomness is to observe random behavior. Computer simulations are useful because they help develop insight into the meaning of random variation. Excel is well suited for simulation and provides both a RAND() function and a **Random Number Generation** tool for such purpose.

## 4.1   Randomness

### Simulation

> **Example 4.1.**(Example 4.1, page 242 in text.)    When you toss a coin, there are only two possible outcomes, heads or tails. Using the RAND() function, simulate tossing a coin 5000 times. For each number of tosses from 1 to 5000 plot the proportion of those tosses that gave a head. Also show on the same graph a horizontal line at the height 0.50.

**Solution.**    The RAND() function produces a number uniformly distributed on the interval (0,1). This can be converted into integers ,taking the values 0 or 1 with equal probability if this uniform random number is multiplied by 2 and then the integer part is taken. The Excel formula for these operations is = INT(2*RAND()).

1. Enter the formula = INT(2*RAND()) in cell A5 of a new workbook and copy this formula down to cell A5003 by selecting cell A4, then clicking the fill

handle and dragging to cell A5003 to generate 5000 tosses of a fair coin (0 representing tails and 1 representing heads).

2. Enter the value 0 in cell B3 followed by the formula = A4+B3 in cell B4. Copy the formula in cell A4 down to cell A5003. Column B tracks the cumulative number of heads.

3. Enter the number 1 in cell C4 and fill to cell C5003 with successive integers $\{1, 2, \ldots, 5000\}$. This can be achieved efficiently by selecting cell C4 and then choosing **Edit − Fill − Series** from the Menu Bar. Complete the **Series** dialog box with Series in **Columns**, Type **Linear**, and **Step Value** 1, **Stop Value** 5000. Click **OK**. Column C will label the 5000 tosses.

4. Fill cells D4 to D5003 with the value 0.5. This will represent the horizontal line at height 0.5 on the graph.

5. Enter the formula = B4/C4 in cell E4 and copy to cell E5003.

Figure 4.1 shows part of the workbook with the required formulas. Next construct

| | A | B | C | D | E |
|---|---|---|---|---|---|
| 1 | **Formulas Behind Simulation of 1000 Tosses** | | | | |
| 2 | | | | | |
| 3 | | 0 | 0 | 0.5 | 0.0000 |
| 4 | =INT(2*RAND()) | =A4+B3 | 1 | 0.5 | =B4/C4 |
| 5 | =INT(2*RAND()) | =A5+B4 | 2 | 0.5 | =B5/C5 |
| 6 | =INT(2*RAND()) | =A6+B5 | 3 | 0.5 | =B6/C6 |

Figure 4.1: Formulas for Simulating 5000 Tosses of a Fair Coin

a graph of the same results. Click the **ChartWizard** button.

**Users of Excel 5/95**

- In Step 1 enter the data range C4:E5003.
- In Step 2 click the **Line** chart type.
- In Step 3 select Format **1**.
- In Step 4 click the button for Data Series in **Columns**. Enter "1" for Use First 1 Column for Category(X) axis labels and enter "0" for Use First 0 Column for Legend Text.
- In Step 5 click the radio button **No** for Add a legend? and label the chart and X axis as shown in Figure 4.2. Click **Finish**.

**Users of Excel 97/98/2000/2001**

- In Step 1 click the **Line** chart type and the second Chart sub-type **Line**.

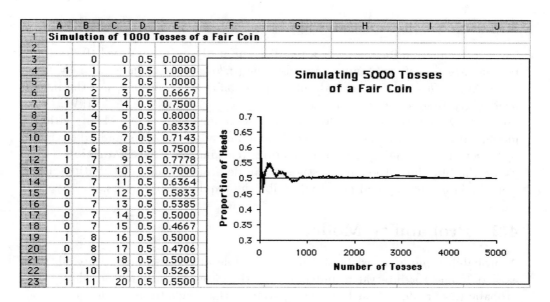

| | A | B | C | D | E | F | G | H | I | J |
|---|---|---|---|---|---|---|---|---|---|---|
| 1 | Simulation of 1000 Tosses of a Fair Coin | | | | | | | | | |
| 2 | | | | | | | | | | |
| 3 | | 0 | 0 | 0.5 | 0.0000 | | | | | |
| 4 | 1 | 1 | 1 | 0.5 | 1.0000 | | | | | |
| 5 | 1 | 2 | 2 | 0.5 | 1.0000 | | | | | |
| 6 | 0 | 2 | 3 | 0.5 | 0.6667 | | | | | |
| 7 | 1 | 3 | 4 | 0.5 | 0.7500 | | | | | |
| 8 | 1 | 4 | 5 | 0.5 | 0.8000 | | | | | |
| 9 | 1 | 5 | 6 | 0.5 | 0.8333 | | | | | |
| 10 | 0 | 5 | 7 | 0.5 | 0.7143 | | | | | |
| 11 | 1 | 6 | 8 | 0.5 | 0.7500 | | | | | |
| 12 | 1 | 7 | 9 | 0.5 | 0.7778 | | | | | |
| 13 | 0 | 7 | 10 | 0.5 | 0.7000 | | | | | |
| 14 | 0 | 7 | 11 | 0.5 | 0.6364 | | | | | |
| 15 | 0 | 7 | 12 | 0.5 | 0.5833 | | | | | |
| 16 | 0 | 7 | 13 | 0.5 | 0.5385 | | | | | |
| 17 | 0 | 7 | 14 | 0.5 | 0.5000 | | | | | |
| 18 | 0 | 7 | 15 | 0.5 | 0.4667 | | | | | |
| 19 | 1 | 8 | 16 | 0.5 | 0.5000 | | | | | |
| 20 | 0 | 8 | 17 | 0.5 | 0.4706 | | | | | |
| 21 | 1 | 9 | 18 | 0.5 | 0.5000 | | | | | |
| 22 | 1 | 10 | 19 | 0.5 | 0.5263 | | | | | |
| 23 | 1 | 11 | 20 | 0.5 | 0.5500 | | | | | |

Figure 4.2: Simulating 5000 Tosses of a Fair Coin

- In Step 2 on the **Data Range** tab enter D4:E5003 for the Data range check the radio button **Series in: Columns**.
- In Step 3 on the **Titles** tab, enter the title and labels of the axes, on the **Axes** tab check radio button **Automatic** for Category (X) axis and check the Value (Y) axis box, on the **Gridlines** tab turn off all gridlines, on the **Legend** tab clear the legend, and finally on the **Data Labels** tab select the radio button **None**.
- In Step 4 embed the graph in the current workbook. Click **Finish**.

Format the X and Y axes as shown in Figure 4.2, for instance by changing the number of categories between tick marks and reorienting the X axis labels. We have also included initial values in row 3 of the worksheet in order to have the labels start at 0. Include data in this row in your data range, otherwise you may find that your $x$-axis starts at 1.

Figure 4.2 shows a segment of the completed workbook with the embedded graph. As previously, you can reevaluate all functions and the graph will dynamically change. From column A you can see the random sequence of heads and tails generated, while column E exhibits the proportions of heads. These are quite variable at first but then settle down, appearing to approach the value 0.5 (shown by the horizontal line). This behavior is known in statistics as a law of large numbers, commonly referred to as the "law of averages."

**Exercise.** Modify the above spreadsheet so that it simulates tossing a "biased" coin.

## Simulating Other Distributions

The Excel function `RAND()` picks a number uniformly on the interval $(0,1)$. Then, to generate a uniform random variable on $(a, b)$ simply change scale by using $a + \text{RAND}() * (b - a)$. To generate other probability distributions, use the inverse probability function $h(a) = \inf\{x : F(x) \geq a\}$. For instance, `NORMINV(RAND(),` Mean, StDev) returns a random normal with mean given by Mean (either a numerical value or a named reference to a numerical value) and standard deviation by StDev. By examining the list of functions (clicking the **Paste Function** button $f_x$ on the Standard Toolbar), a student can determine which distributions Excel can simulate this way and how to describe the required parameters.

## 4.2   Probability Models

A probability model consists of a list of possible outcomes and a probability for each outcome (or interval of outcomes, in the case of continuous models). The probabilities are determined by the experiment that leads to the occurrence of one or more of the outcomes in the specified list.

Excel provides many distributions that may be constructed in a common fashion with the **Formula Palette** or the **Function Wizard**. The meaning of the required parameters is available online through Excel's help feature. Because of its prominence, the normal distribution was already discussed in Chapter 1. In this chapter and in the next we shall introduce important probability models and simulate from them.

## Uniform

A uniform random number is one whose values are spread out uniformly across the interval from 0 to 1. Its density curve has height 1 over the interval 0 to 1.

> **Example 4.2.**   (Based on Example 4.7, page 257 in text.) Let $Y1$ be a uniform random number between 0 and 1. Use Excel to generate 1000 random uniform numbers, and from your simulations estimate the following probabilities and compare them with the theoretical values.
> (a) $P(0.3 \leq Y \leq 0.7)$
> (b) $P(Y \leq 0.5)$
> (c) $P(Y > 0.7)$
> (d) $P(Y \leq 0.5 \text{ or } Y > 0.8)$

**Solution.**   The theoretical values are 0.400, 0.500, 0.200, and 0.7 respectively. Use `RAND()` to generate 1000 uniform random variables in a column and construct a histogram with bin intervals of width 0.10, beginning at 0 and ending at 1. Figure 4.3 shows a portion of a worksheet where this has been done. The frequencies shown are the number of times the random number generator produced a number

| | A | B | C | D | E | F | G |
|---|---|---|---|---|---|---|---|
| 1 | **Uniform Random Numbers** | | | | | | |
| 2 | | | | | | | |
| 3 | Y1 | n | Y2 | Y1+Y2 | Bin | Bin | Frequency |
| 4 | 0.77967 | 1 | 0.49868 | 1.27834 | 0.10 | 0.1 | 102 |
| 5 | 0.53383 | 2 | 0.93028 | 1.46411 | 0.20 | 0.2 | 95 |
| 6 | 0.29443 | 3 | 0.51077 | 0.80519 | 0.30 | 0.3 | 85 |
| 7 | 0.43453 | 4 | 0.68530 | 1.11982 | 0.40 | 0.4 | 104 |
| 8 | 0.49917 | 5 | 0.60503 | 1.10421 | 0.50 | 0.5 | 107 |
| 9 | 0.21515 | 6 | 0.21059 | 0.42574 | 0.60 | 0.6 | 96 |
| 10 | 0.43022 | 7 | 0.37694 | 0.80716 | 0.70 | 0.7 | 99 |
| 11 | 0.16801 | 8 | 0.20419 | 0.37220 | 0.80 | 0.8 | 91 |
| 12 | 0.60948 | 9 | 0.95073 | 1.56021 | 0.90 | 0.9 | 112 |
| 13 | 0.30827 | 10 | 0.73422 | 1.04249 | 1.00 | 1 | 109 |
| 14 | 0.94542 | 11 | 0.18256 | 1.12798 | | | |
| 15 | 0.10970 | 12 | 0.55314 | 0.66284 | | | |
| 16 | 0.63897 | 13 | 0.50625 | 1.14521 | | | |
| 17 | 0.04368 | 14 | 0.17084 | 0.21452 | | | |
| 18 | 0.03094 | 15 | 0.09285 | 0.12379 | | | |
| 19 | 0.10825 | 16 | 0.37989 | 0.48814 | | | |

Figure 4.3: Worksheet for Simulating 1000 Uniform Random Variables

$Y1$ in the specified interval. (This worksheet is also used in Example 4.2, hence the additional columns involving $Y2$.) The values listed under the heading *Bin* are the right endpoints of the intervals. We count the number of observations in the relevant intervals and divide by 1000 to convert to a probability and obtain:

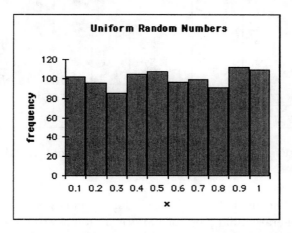

Figure 4.4: Histogram of Simulation of 1000 Uniform Random Variables

(a) $P(0.3 \leq Y \leq 0.7) = (104+107+96+99)/1000 = 0.406$
(b) $P(Y \leq 0.5) = (102+95+85+104+107)/1000 = 0.493$
(c) $P(Y > 0.7) = (112 + 109)/1000 = 0.221$
(d) $P(Y \leq 0.5 \text{ or } Y > 0.8) = 0.493 + 0.221 = 0.714$

## Triangular—Adding Random Numbers

**Example 4.3.** (Based on Exercise 4.54, page 317 in text.) Generate
two random numbers between 0 and 1 and take their sum. Repeat
this 1000 times and construct a histogram of the results. The sum of
two uniform continuous [0,1] valued random variables has a triangular
distribution and histogram should have a triangular shape. From the
simulations estimate the following probabilities.

(a) P($0 \leq X \leq 0.5$)
(b) P($0 \leq X \leq 1.0$)

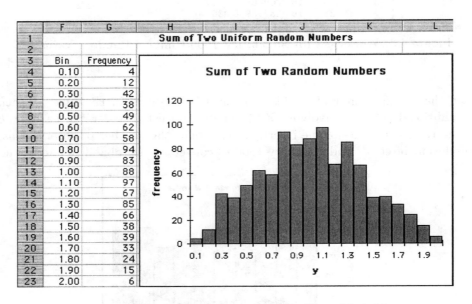

Figure 4.5: Simulating Triangular Random Variables

**Solution.** It is known that the idealized density curve of Y is a triangle. Again
use RAND() to generate 1000 pairs of uniform random variables {$Y1, Y2$} in two
columns, add the columns, and construct a histogram with bin intervals of width
0.10 beginning at 0 and ending at 2. (See earlier worksheet show in Figure 4.3
Figure 4.5 shows the sample output from a workbook where this has been done.
The frequencies shown are the number of times the random number generator
produced a number in the specified interval.

(a) P($0 \leq X \leq 0.5$) = $(4 + 12 + 42 + 38 + 49)/1000) = 0.145$
(b) P($0 \leq X \leq 1.0$) = $(4+12+42+38+49+62+58+94+93+88)/1000 = 0.540$
The corresponding theoretical values are 0.125 and 0.500, respectively.

## 4.3   Random Variables

A random variable $X$ is a variable whose value is a numerical outcome of a random phenomenon. It is completely prescribed by its probability distribution, which tells what values $X$ can take and what probabilities are assigned to these values. Random variables can be discrete or continuous and Excel includes the distributions of many random variables in its library, including the beta, binomial, chi-square, exponential, F, gamma, lognormal, negative binomial, normal, Student t, uniform, and Weibull, as well as their inverses. Here we examine a few of these.

### Binomial

This will be treated in detail in Chapter 5.

### Hypergeometric

HYPERGEOMDIST$(x, n, M, N,)$ provides probabilities for an experiment in which a simple random sample of size $n$ is taken from a finite population of $N$ individuals of which $M$ are in a so-called "preferred category" called "success" or "1," while the remaining $N - M$ are deemed "failure" or "0." The function returns the probability of $x$ successes in the sample of size $n$.

> **Example 4.4.**   (Lotto 6/49) Suppose a box contains 49 balls in which one and only one ball is marked with an integer taken from $\{1, 2, \ldots, 49\}$. The balls are identical otherwise. Suppose that the balls numbered $\{1, 2, 3, 4, 5, 6\}$ are considered "successes." If an SRS of six balls is taken at random (without replacement), what is the probability that the sample contains $k$ successes (for $k = 0, 1, 2, 3, 4, 5, 6$)?

|    | A | B | C |
|----|---|---|---|
| 1  | \multicolumn{3}{c}{**Lotto 6/49 Probabilities**} | | |
| 2  |   |   |   |
| 3  | k | P(X=k) value | P(K=k) formula |
| 4  | 0 | 0.4359650 | =HYPGEOMDIST(A4,6,6,49) |
| 5  | 1 | 0.4130195 | =HYPGEOMDIST(A5,6,6,49) |
| 6  | 2 | 0.1323780 | =HYPGEOMDIST(A6,6,6,49) |
| 7  | 3 | 0.0176504 | =HYPGEOMDIST(A7,6,6,49) |
| 8  | 4 | 0.0009686 | =HYPGEOMDIST(A8,6,6,49) |
| 9  | 5 | 0.0000184 | =HYPGEOMDIST(A9,6,6,49) |
| 10 | 6 | 0.0000001 | =HYPGEOMDIST(A10,6,6,49) |

Figure 4.6: Calculating Lotto Probabilities

**Solution.**   The answer is provided in Figure 4.6 where column B gives the probabilities and column C, the corresponding Excel formulas.

## Student *t* Distribution

The normal distribution was discussed in detail in Chapter 1. A related distribution is the Student *t* found in 1908 by W. Gossett (who used the pseudonym "Student" when he published his paper). It arises as the distribution of the *Studentized* score (similar to a standardized score)

$$t = \frac{\bar{x} - \mu}{s/\sqrt{n}}$$

where $\bar{x}, s$ are the sample mean and sample standard deviation from a sample of size $n$ taken from a $N(\mu, \sigma)$ population. It is determined by a single parameter called the degrees of freedom $\nu$. In the above ratio, $\nu$ takes the value $n - 1$.

Excel has an unusual definition of the c.d.f. and inverse c.d.f. for the *t* distribution. The Excel function TDIST returns the tail of the distribution, that is, if $t(\nu)$ is a random variable with a *t* distribution on $\nu$ d.f., then

$$\text{TDIST}(x, \nu, 1) = P[t(\nu) > x]$$

and

$$\text{TDIST}(x, \nu, 2) = 2P[t(\nu) > x]$$

The argument $x$ in TDIST must be positive. Thus the c.d.f. takes a rather complicated expression using the logical IF function:

$$P(t(\nu) \le x) = \text{IF}(x < 0, \text{TDIST}(\text{ABS}(x), \nu, 1), 1 - \text{TDIST}(x, \nu, 1))$$

where $\text{ABS}(x) = |x|$, and similarly

$$P(|t(\nu)| \le x) = 1 - \text{TDIST}(x, \nu, 2)$$

The inverse function TINV is defined by

$$P[t(\nu) > \text{TINV}(\alpha, \nu)] = \frac{\alpha}{2}$$

so $\text{TINV}(\alpha, \nu)$ is the critical value for a two-sided significance test at level $\alpha$ of a normal mean (to be discussed in Chapter 7).

> **Exercise.** Using NORMSDIST and TDIST, graph on the same figure and to the same scale the densities of a $N(0, 1)$ and the Student *t* distribution on 4 d.f. (Figure 4.7).

## Simulating Random Variables Using Random Number Generation

In addition to the function RAND(), Excel has a **Random Number Generation** tool built into the **Analysis ToolPak** that provides an alternative and more systematic approach to simulation.

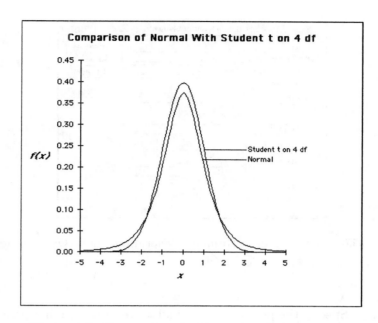

Figure 4.7: Comparing the Normal and the *t* Curves

The **Random Number Generation** tool creates columns of random numbers, as specified by the user, from any of six probability models (uniform, normal, Bernoulli, binomial, Poisson, discrete) as well as having an option for patterned that creates not random data but rather data according to a specified pattern.

All options are invoked from a common dialog box (as in Figure 4.8 for discrete) following the choice **Tools – Data Analysis – Random Number Generation** from the Menu Bar. Select the distribution of interest using the drop-down arrow and the Parameters sub-box will automatically change, prompting the input of parameters.

**Number of Variables.** Enter the number of columns of random variables. The default is all columns.

**Number of Random Numbers.** Enter the number of rows (cases) of random variables.

**Distributions.** Use the drop-down arrow to open a list of choices with requested parameters.

> **Poisson.** Upper and lower limits
> **Normal.** $\mu, \sigma$
> **Bernoulli.** $p$ = probability of success
> Excel unfortunately refers to this as a $p$ Value.
> **Binomial.** $p, n$

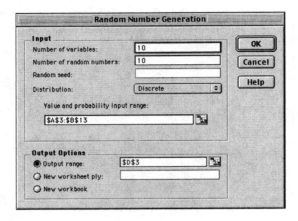

Figure 4.8: Random Number Generation Tool—Discrete

**Poisson.**     $\lambda$

**Discrete.**    Specify the possible values and their corresponding probabilities. Before using this option enter the values and probabilities in adjacent columns in the workbook.

**Patterned.**   This option creates data according to a prescribed pattern of values repeated in specified steps. This is useful if a linear array of data needs to be coded using another variable.

**Example 4.5.**     To generate 100 tosses of a pair of fair dice enter $\{2, 3, \ldots, 11\}$ into cells A3:A13 and enter

$$\{1/36, 2/36, \ldots, 6/36, 5/36, \ldots, 2/36, 1/36\}$$

into cells B3:B12 (Figure 4.9). Excel may interpret the value 1/36 as a date Jan 1936. If this happens then format the cells by choosing

| | A | B | C | D | E | F | G | H | I | J | K | L | M |
|---|---|---|---|---|---|---|---|---|---|---|---|---|---|
| 1 | | Simulation of a Pair of Fair Dice | | | | | | | | | | | |
| 2 | k | P(X=k) | | | | | | | | | | | |
| 3 | 2 | 0.0277778 | | 2 | 8 | 6 | 8 | 4 | 7 | 7 | 8 | 6 | 5 |
| 4 | 3 | 0.0555556 | | 6 | 9 | 5 | 6 | 8 | 9 | 12 | 9 | 7 | 7 |
| 5 | 4 | 0.0833333 | | 8 | 5 | 8 | 8 | 12 | 6 | 7 | 8 | 9 | 5 |
| 6 | 5 | 0.1111111 | | 2 | 9 | 3 | 8 | 5 | 7 | 7 | 6 | 5 | 8 |
| 7 | 6 | 0.1388889 | | 8 | 6 | 11 | 10 | 6 | 8 | 4 | 8 | 7 | 8 |
| 8 | 7 | 0.1666667 | | 10 | 6 | 9 | 8 | 2 | 6 | 4 | 7 | 6 | 4 |
| 9 | 8 | 0.1388889 | | 9 | 3 | 4 | 5 | 10 | 7 | 7 | 6 | 4 | 7 |
| 10 | 9 | 0.1111111 | | 6 | 6 | 4 | 11 | 9 | 7 | 7 | 2 | 11 | 4 |
| 11 | 10 | 0.0833333 | | 6 | 9 | 8 | 9 | 5 | 8 | 3 | 12 | 3 | 4 |
| 12 | 11 | 0.0555556 | | 12 | 8 | 3 | 5 | 8 | 4 | 10 | 7 | 10 | 7 |
| 13 | 12 | 0.0277778 | | | | | | | | | | | |

Figure 4.9: Simulating a Pair of Fair Dice

**Format – Cells** from the Menu Bar and selecting **Number**. Then choose **Tools – Data Analysis – Random Number Generation** from the Menu Bar and complete as in Figure 4.8. The output will appear in cells D1:M10. Because these numbers are random, your output will of course be different.

## 4.4   The Sampling Distribution of a Sample Mean

A statistic based on a random sample will take different values for different samples. The values it takes in repeated samples do not vary haphazardly from sample to sample but follow a regular pattern called its sampling distribution. For instance in Examples 4.2 and 4.3, we took repeated samples of size 1 and saw that the empirical histograms in Figure 4.4 and Figure 4.5 approximated the true theoretical histograms.

In this section, we examine not a sample of size one, but rather the sample mean $\overline{x}_n$ of a sample of size $n$. Simulation followed by a histogram of the results provides an insightful view of the **Central Limit Theorem**.

### Law of Large Numbers

In Example 4.1 we simulated tossing a fair coin 5000 times and recorded the cumulative proportion of heads (the sample mean) after each toss. This gave us a record of the averages $\overline{x}_n$ for $n = 1, 2, \ldots, 5000$. We saw that the average tends to settle down to the value 0.50. This long-run frequency behavior illustrates the **Law of Large Numbers**.

> Draw independent observations at random from any population with finite mean $\mu$. As the number of observations drawn increases, the mean $\overline{x}$ of the observed values gets closer and closer to the mean $\mu$ of the population.

### The Mean and Standard Deviation of $\overline{x}$

The mean and standard deviation of $\overline{x}$ are also related to the mean and standard deviation of the population. The results are: mean of $\overline{x} = \mu$ and standard deviation of $\overline{x} = \sigma/\sqrt{n}$, where $\mu$ and $\sigma$ are the mean and standard deviation of the population, respectively.

### Sampling Distribution of a Normal Sample Mean

If a population has the $N(\mu, \sigma)$ distribution, then the sample mean $\overline{x}$ of $n$ independent observations has the $N(\mu, \sigma/\sqrt{n})$ distribution. We illustrate this important fact using simulation.

**Example 4.6.**    (Example 4.21–4.23 pages 293–303 in text.) Sulfur compounds such as dimethyl sulfide (DMS) are sometimes present in wine. DMS causes "off odors" in wine, so winemakers want to know the odor threshold, the lowest concentration of DMS that the human nose can detect. Different people have different thresholds, so we start by asking about the mean threshold $\mu$ in the population of all adults. The number $\mu$ is a *parameter* that describes this population. To estimate $\mu$, we present tasters with both natural wine and the same wine spiked with DMS at different concentrations to find the lowest concentration at which they can identify the spiked wine. Here are the odor thresholds (measured in micrograms of DMS per liter of wine) for 10 randomly chosen subjects:

$$28 \; 40 \; 28 \; 33 \; 20 \; 3 \; 129 \; 27 \; 17 \; 21$$

The sample mean of these 10 numbers is $\bar{x} = 27.4$ The parameter $\mu$ is a fixed, unknown number and it seems reasonable to use $\bar{x}$ to estimate $\mu$. But $\bar{x}$ will vary from sample to sample so we can't expect $\bar{x}$ to be near $\mu$ all the time. If we chose a different sample of 10 subjects then we would get a different value of $\bar{x}$. Extensive studies have found that the DMS odor threshold of adults follows roughly a normal distribution with mean $\mu = 25$ micrograms per liter and standard deviation $\sigma = 7$ micrograms per liter. Take 1000 samples of size 10. For each sample find $\bar{x}$ the sample mean and then make a histogram of these 1000 values.

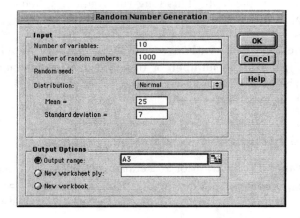

Figure 4.10: Random Number Generation Tool – Normal

**Solution.**

1. Choose **Tools – Data Analysis – Random Number Generation** from the Menu Bar and complete the dialog box as in Figure 4.10. The output will

appear in cells A3:J1002 (10 columns by 1000 rows). Each row represents a sample of size 10.

2. Enter the formula = `AVERAGE`(A3:J3) in cell K3 to compute the average of the first sample of size four.

3. Select cell K3, click the fill handle, and drag down to cell K1002 to obtain the averages of all the rows. See Figure 4.11 showing the first seven samples. Because these numbers are random, your output will of course be different.

4. From the Menu Bar choose **Tools − Data Analysis** and scroll to **Histogram** and construct a histogram of the data. See Figure 4.12.

| | A | B | C | D | E | F | G | H | I | J | K |
|---|---|---|---|---|---|---|---|---|---|---|---|
| 1 | Sum of 10 Normal Random Variables | | | | | | | | | | |
| 2 | | | | | | | | | | | |
| 3 | 12.845 | 19.261 | 30.999 | 17.543 | 21.822 | 23.399 | 24.207 | 26.528 | 20.657 | 21.226 | 21.849 |
| 4 | 26.788 | 24.658 | 23.261 | 25.003 | 23.272 | 24.115 | 31.473 | 26.214 | 28.371 | 17.199 | 25.035 |
| 5 | 33.669 | 28.681 | 28.678 | 20.911 | 26.808 | 26.646 | 24.334 | 25.482 | 21.488 | 21.047 | 25.774 |
| 6 | 22.058 | 21.025 | 27.947 | 26.715 | 15.537 | 24.991 | 28.966 | 19.112 | 26.708 | 25.152 | 23.821 |
| 7 | 27.996 | 25.786 | 24.842 | 14.356 | 25.984 | 26.865 | 18.778 | 20.431 | 13.106 | 28.323 | 22.647 |
| 8 | 18.091 | 25.692 | 24.541 | 11.894 | 26.164 | 21.968 | 19.545 | 23.493 | 29.451 | 33.636 | 23.448 |
| 9 | 43.398 | 14.449 | 16.291 | 21.091 | 24.120 | 18.500 | 19.726 | 29.149 | 31.402 | 22.831 | 24.096 |

Figure 4.11: Samples of size 10 with Sample Means from $N(25, 7)$

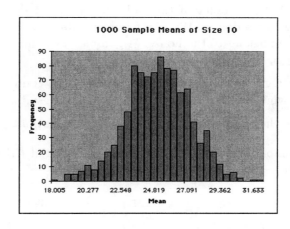

Figure 4.12: Histogram − 1000 Samples Means of size 10 from $N(25, 7)$

The histogram is centered near the theoretical mean of 10. Type the Excel formula = `AVERAGE(K3:K1002)` into a blank cell to find the average. It appears to be normally shaped and to extend from about 18 to 32. Applying the 65-95-99.7 % rule and taking the difference 32-18 = 14 as six standard deviation units, we divide 14 by 6 to get an estimated standard deviation of 2.33, which compares very favorably with the theoretical value of $7/\sqrt{10} = 2.21$.

## Simulating the Central Limit Theorem – Non-Normal Population

We have learned that when the underlying population is normal, then the sample mean $\bar{x}$ has an exact normal distribution. When the distribution is not normal, then the distribution of $\bar{x}$ depends on the population and is not normal. But when the sample size gets large, a remarkable stability occurs. This is called the **Central Limit Theorem.**

> Draw an SRS of size $n$ from any population with mean $\mu$ and finite standard deviation $\sigma$. When $n$ is large, the sampling distribution of the sample mean $\bar{x}$ satisfies:
>
> $$\bar{x} \text{ is approximately } N(\mu, \sigma/\sqrt{n})$$

We provide a simulation demonstration of the Central Limit Theorem in the next example.

> **Example 4.7.** (Based on Exercise 4.91, page 306 in text.) A roulette wheel has 38 slots – 18 are black, 18 are red, and 2 are green. When the wheel is spun, a ball is equally likely to come to rest in any of the slots. Gamblers can place a number of different bets in roulette. One of the simplest wagers chooses red or black. A bet of one dollar on red will pay off an additional dollar if the ball lands in a red slot. Otherwise the player loses his dollar. When a gambler bets on red or black, the two green slots belong to the house. A gambler's winnings on a \$1 bet are either \$1 or −\$1.
>
> (a) Simulate a gambler's winnings after 50 bets and compare the gambler's mean winnings per bet with the theoretical results.
>
> (b) Compare the results with the normal approximation.

**Solution.** The number of wins after 50 bets $X$ is a binomial $B(50, 10/38)$ random variable with

$$\text{mean} \qquad \mu_X = 50\left(\frac{18}{38}\right) = 23.684$$

$$\text{standard deviation} \qquad \sigma_X = \sqrt{50\left(\frac{18}{38}\right)\left(\frac{20}{38}\right)} = 3.5306$$

The proportion of wins after 50 bets is $\hat{p} = X/50$ with

$$\text{mean} \qquad \mu_{\hat{p}} = \frac{18}{38} = 0.4737$$

$$\text{standard deviation} \qquad \sigma_{\hat{p}} = \sqrt{\left(\frac{18}{38}\right)\left(\frac{20}{38}\right) \Big/ 50} = 0.0706$$

The gambler either wins \$1 or loses \$1. His average winnings per game, denoted by $\bar{w}$, are therefore $1 \times$ (the proportion of times that $X = 1$) minus $1 \times$ (the proportion of times that $X = -1$), that is $\bar{w} = \hat{p}(1) + (1 - \hat{p})(-1) = 2\hat{p} - 1$. By the rules for means and standard deviations

$$\text{mean} \qquad \mu_{\bar{w}} = 2\mu_{\hat{p}} - 1 = -0.0527$$
$$\text{standard deviation} \qquad \sigma_{\bar{w}} = 2\sigma_{\hat{p}} = 0.14123$$

By the **Central Limit Theorem** $\hat{p}$ is approximately normal, and therefore

$$\bar{w} \text{ is approximately } N(-0.0527, 0.14123).$$

We can simulate a binomial random variable $X$, convert it first to $\hat{p} = \frac{X}{n}$ and then to $\bar{w} = 2\hat{p} - 1$, after which we construct a histogram of the simulation results.

The following steps, referring to Figure 4.13, show how to develop a workbook to simulate 500 replications of 50 games. We will use the RAND function, which *links* the output to a histogram.

| | A | B | C | D | E |
|---|---|---|---|---|---|
| 1 | \multicolumn{5}{c}{**Demonstrating the Central Limit Theorem by Simulation**} | | | | |
| 2 | | | | | |
| 3 | true mean = | -0.0527 | simulated mean = | -0.0556 | |
| 4 | true st_dev = | 0.14123 | simulated st_dev = | 0.14120 | |
| 5 | | | | | |
| 6 | Formula entered in column B | | =2*CRITBINOM(50,18/38,RAND())/50 -1 | | |
| 7 | Simulation | Average per Game | | | |
| 8 | 1 | 0.040 | Bin Formulas | Bin | Freq. |
| 9 | 2 | -0.200 | =-0.0527-3.5*0.14123 | -0.55 | 0 |
| 10 | 3 | -0.040 | =-0.0527-3*0.14123 | -0.48 | 1 |
| 11 | 4 | -0.080 | =-0.0527-2.5*0.14123 | -0.41 | 0 |
| 12 | 5 | -0.120 | =-0.0527-2*0.14123 | -0.34 | 9 |
| 13 | 6 | -0.280 | =-0.0527-1.5*0.14123 | -0.26 | 26 |
| 14 | 7 | 0.120 | =-0.0527-0.14123 | -0.19 | 57 |
| 15 | 8 | 0.000 | =-0.0527-0.5*0.14123 | -0.12 | 54 |
| 16 | 9 | -0.040 | =-0.0527 | -0.05 | 86 |
| 17 | 10 | -0.160 | =-0.0527+0.5*0.14123 | 0.02 | 119 |
| 18 | 11 | -0.080 | =-0.0527+0.14123 | 0.09 | 85 |
| 19 | 12 | -0.160 | =-0.0527+1.5*0.14123 | 0.16 | 27 |
| 20 | 13 | 0.040 | =-0.0527+2*0.14123 | 0.23 | 26 |
| 21 | 14 | -0.120 | =-0.0527+2.5*0.14123 | 0.30 | 7 |
| 22 | 15 | 0.160 | =-0.0527+3*0.14123 | 0.37 | 1 |
| 23 | 16 | 0.040 | =-0.0527+3.5*0.14123 | 0.44 | 2 |
| 24 | 17 | -0.040 | | | 500 |

Figure 4.13: Simulating the Central Limit Theorem

Bin intervals for the histogram will be located at multiples of the standard deviation from the mean.

1. Prepare a new workbook by entering "Simulating the Central Limit Theorem" in cell A1 and centering the heading across A1:E1. Enter "Simulation" in A7 and "Average per Game" in B7.

2. Enter the values $1, 2, \ldots, 500$ in cells A8:A507 as follows: Enter "1" in A8. Select A8 and choose **Edit − Fill − Series...** from the Menu Bar. In

the **Series** dialog box, check Series in **Columns** and Type **Linear**. Clear the **Trend** box and type "1" and "500" for the **Step** and **Stop** values, respectively.

3. Simulate 50 games. In cell B8 enter $= 2*$CRITBINOM$(50,18/38, $RAND$())/50$-1 to generate the random variable $\bar{w}$. We have shown the formula on line 6 beginning in column C. Select B8, click the fill handle at the lower right corner of B8 and drag down to cell B507. Cells B8:B507 are now filled with 500 replications of the gambler's average net gain per game after each simulated 50 games.

4. Next we prepare the simulations for output into a histogram. Enter the labels "Bin" in C8 and "Freq." in D8. The bin endpoints for the histogram appear in cells D9:D22 and the corresponding formulas behind the values are shown in cells C9:C23. These are based on theoretical true mean and standard deviation, and the bin endpoints are expressed in simple multiples of the standard deviation from the mean. Enter $= -0.0527 - 3.5 * 0.14123$ in D9, $= -0.0527 - 3.0 * 0.14123$ in D10, and so on. Refer to Figure 4.13 where we have shown in cells C9:C23 the formulas to be entered in D9:D23. Note that you do not require a column C in your own worksheet.

5. Select E9:E23. Then type $=$ FREQUENCY$($B8:B507,D9:D23$)$ in the entry area of the **Formula Bar**. Hold down the **Shift** and **Control** keys (either **Macintosh** or **Windows**) and press **enter/return** to **array-enter** the formula. The formula will appear **surrounded by braces** { } in the **Formula Bar**, and the bin frequencies will appear in cells E9:E23. Select cells D9:D23 and then complete the sequence of steps in the **ChartWizard** as discussed previously. The resulting histogram appears in Figure 4.14.

6. Note that we have also located the sample mean $\bar{w}$ and sample standard deviation $s_{\bar{w}}$ on the worksheet for comparison with the true values. These will change whenever the worksheet is re-evaluated. The formula behind cell D3 is =AVERAGE(B8:B507) and behind cell D4 it is $=$ SQRT$((1$-D3*D3$)/50)$.

## Excel Output

The sample mean and sample standard deviation appear in D3:D4, and the population mean and standard deviation appear in B3:B4 for comparison purposes. For the simulation shown

$$\bar{w} = -0.0556 \qquad \mu_{\bar{w}} = -0.0527$$
$$s_{\bar{w}} = \phantom{-}0.14120 \qquad \sigma_{\bar{w}} = \phantom{-}0.14123$$

The table of frequency counts appears in E9:E23 with the corresponding histogram in Figure 4.14. The histogram appears normal shaped with no unusual features.

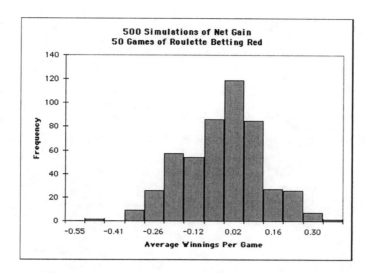

Figure 4.14: Histogram of 500 Simulations of 50 Roulette Games

Recalling that the bin entries are the right endpoints of the bin interval, we can determine the proportion of counts within 1, 2, and 3 standard deviation units of the mean. The simulation results are alarmingly good.

|  | Actual | Theoretical |
|---|---|---|
| Within 1 $\sigma$ | .688 | .683 |
| Within 2 $\sigma$ | .960 | .954 |
| Within 3 $\sigma$ | .994 | .997 |

# Chapter 5

# Probability Theory

Probability models are theoretical descriptions of variation and can be used to describe diverse phenomena. An important class of models arises when the data are counts of some variable. This leads to the binomial model for sample counts and sample proportions. In Chapter 4 we used simulation as an approach to understanding variation. This is continued here.

## 5.1   General Probability Rules

Many probability models are derived from independent outcomes. The multiplication rule says that the probability of two independent events occurring is the product of their individual probabilities.

> **Example 5.1.**   (Based on Exercise 5.16 – 5.18 page 328 in text.) All human blood can be "ABO-typed" as one of O, A, B, or AB, but the distribution of the types varies a bit among groups of people. Here is the distribution of blood types for a randomly chosen person in the United States.
>
> | Blood type | O | A | B | AB |
> |---|---|---|---|---|
> | U.S. Probability | 0.45 | 0.40 | 0.11 | 0.04 |
>
> Choose a married couple at random. It is reasonable to assume that the blood types of husband and wife are independent and follow this distribution.
>
> (a) **Is transfusion safe?** Someone with type B blood can safely receive transfusions only from persons with type B or type O blood. What is the probability that the husband of a woman with type B blood is an acceptable blood donor for her?
>
> (b) **Same type?** What is the probability that a wife and husband share the same blood type?

| | A | B | C | D | E | F | G | H | I |
|---|---|---|---|---|---|---|---|---|---|
| 1 | Blood Types | | | | | | | | |
| 2 | | | | | | Wife = | Husband = | | |
| 3 | | | | Wife | Husband | Type B | Type B or O | =SUM(F4:F1003) | 109 |
| 4 | Type | k | P(X=k) | 1 | 2 | 0 | 0 | =SUM(G4:G1003) | 61 |
| 5 | O | 1 | 0.45 | 1 | 2 | 0 | 0 | =I4/I3 | 0.5596 |
| 6 | A | 2 | 0.40 | 1 | 2 | 0 | 0 | | |
| 7 | B | 3 | 0.11 | 2 | 2 | 0 | 0 | | |
| 8 | AB | 4 | 0.04 | 1 | 1 | 0 | 0 | | |
| 9 | | | | 1 | 2 | 0 | 0 | | |
| 10 | | | | 1 | 1 | 0 | 0 | | |
| 11 | | | | 2 | 2 | 0 | 0 | | |
| 12 | | | | 4 | 2 | 0 | 0 | | |
| 13 | | | | 2 | 2 | 0 | 0 | | |
| 14 | | | | 2 | 1 | 0 | 0 | | |
| 15 | | | | 2 | 2 | 0 | 0 | | |
| 16 | | | | 4 | 1 | 0 | 0 | | |

Figure 5.1: Simulating Blood Types

**Solution.**

(a) This problem can easily be solved without Excel. We are interested in the conditional probability

P[husband is of type O or B | wife is of type O].

Because of independence this is the same as

P[husband is of type O or B] = 0.45 + 0.11 = 0.56.

We will now solve this in Excel. There are several reasons for doing this. First, it will introduce use of the logical functions IF, AND, and OR. Second, it uses the frequency interpretation of probability and conditional probability, and confirms the accuracy of such an interpretation. Third, the Excel approach via simulation generalizes to more complicated situations where an exact analytic solutions is not feasible.

Enter the probabilities of the various blood types into an Excel worksheet (see Figure 5.1, where the types have been entered into cells A5:A8, with numerical codes in B5:B8 and the probabilities in C5:C8). Next we generate 1000 husband-wife pairs. Choose **Tools – Data Analysis – Random Number Generation** from the Menu Bar and complete as in Figure 5.2. The output will appear in cells D4:E1003. Next use the conditional IF function to determine when the wife is type B (code = 3). Enter =IF(D4 = 3, 1,0) in cell D1 and fill to cell D1003. This function fills column D with the integer 1 if the wife is type B, otherwise with the integer 0 if the wife is not of type B. Then enter =IF(AND(D4 = 3, OR(E4 =1, E4=3)), 1,0). This function fills column E with the integer 1 if the husband is of type O or B and at the same time the wife is type B, otherwise with the integer 0. Finally, in three empty cells (we have used I3:I5), use Excel to calculate column F sum (=SUM(F4:F1003)), column G sum (=SUM(G4:G1003)), and the ratio of these numbers (=I4/I). What we are calculating is the desired conditional probability. The value on the worksheet is 0.5596, which compares favorably with the theoretical value 0.56.

(b) can be done in the same way and the solution is not shown.

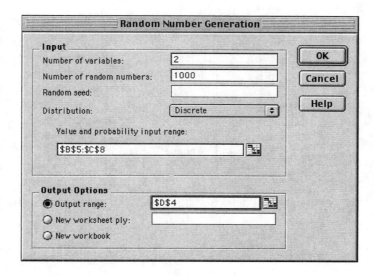

Figure 5.2: Random Number Generation Tool

## 5.2    The Binomial Distributions

A binomial distribution is associated with an experiment comprised of $n$ independent trials each of which has the same success probability $p$. The random variable $X$ counts the number of successes.

It is known that

$$P(X = k) \;\;=\;\; \binom{n}{k} p^k (1-p)^{n-k} \qquad k = 0, 1, 2, \ldots, n$$

$$\text{mean} \;\;=\;\; \mu_X = np$$

and

$$\text{standard deviation} \;=\; \sigma_X = \sqrt{np(1-p)} \;.$$

The corresponding Excel function is $\texttt{BINOMDIST}(k, n, p, \text{cumulative})$. If the parameter cumulative is set to "false," Excel returns the probabilities $P(X = k)$, while if it is set to "true" Excel returns the cumulative probabilities $P(X \leq k)$.

**Example 5.2.**    (Example 5.10, page 335 in text.)  This example refers back to Case 5.1 on page 331. A quality engineer inspects an SRS of 10 switches from a large shipment of which 10% fail to conform. The count X of non-conforming switches in the sample of size 10 has a Binomial distribution with n = 10 and p = 0.10. Construct a table of these binomial probabilities and their cumulative probabilities.

| | A | B | C |
|---|---|---|---|
| 1 | **Binomial Probabilities** | | |
| 2 | | | |
| 3 | k | P(X=k) | P(X<=k) |
| 4 | 0 | 0.34868 | 0.34868 |
| 5 | 1 | 0.38742 | 0.73610 |
| 6 | 2 | 0.19371 | 0.92981 |
| 7 | 3 | 0.05740 | 0.98720 |
| 8 | 4 | 0.01116 | 0.99837 |
| 9 | 5 | 0.00149 | 0.99985 |
| 10 | 6 | 0.00014 | 0.99999 |
| 11 | 7 | 0.00001 | 1.00000 |
| 12 | 8 | 0.00000 | 1.00000 |
| 13 | 9 | 0.00000 | 1.00000 |
| 14 | 10 | 0.00000 | 1.00000 |

Figure 5.3: Binomial Probabilities for $n = 10$ and $p = 0.10$

**Solution.**

1. Enter the label $k$ in cell A3 and the label $P(X = k)$ in cell B3 of a new workbook. In A4:A14 enter the values $\{0, 1, 2, \ldots, 10\}$.

2. Activate cell B4. Using either the **Function Wizard** or the **Formula Palette**, construct the binomial function by selecting **Statistical** for Function Category and BINOMDIST for Function Name.

3. Input the following into the dialog box.

    | | |
    |---|---|
    | number_s | Enter the cell address A4. |
    | trials | Enter the value 10. |
    | probability_s | Enter the value 0.1. |
    | cumulative | Enter the value 0. |

    Your completed formula should look like =BINOMDIST(A4,10,0.1,0). Click **Finish** or **OK**.

4. Activate cell B4, click the fill handle in the lower right corner, and drag to cell B14 to fill the column with individual binomial probabilities (Figure 5.3).

5. Next label cell C3 as $P(X <= k)$ and repeat Steps 2, 3, and 4. Activate C4 instead of B4 in Steps 2 and 4 and enter the value 1 for the cumulative distribution in Step 3.

The resulting table of individual and cumulative binomial probabilities appears in Figure 5.3.

## Binomial Distribution Histogram

We can quickly construct a histogram using the **ChartWizard** displaying the binomial probabilities just calculated. As the procedure is identical to earlier constructions of charts, we omit the details. This histogram appears in Figure 5.4.

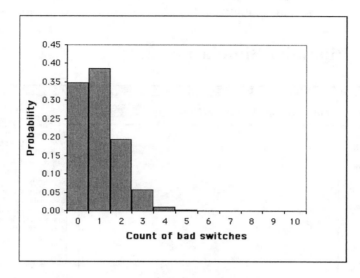

Figure 5.4: Binomial Histogram for $n = 10$ and $p = 0.10$

## Normal Approximation to the Binomial

When $n$ is large, the distribution of the Binomial distribution with success probability $p$ is approximately Normal $N(np, \sqrt{np(1-p)})$. This is a special case of the Central Limit Theorem discussed in Section 4.4 where an example using Excel and involving the Binomial distribution was given. The reader is referred to that section.

## Inverse Cumulative Binomial

CRITBINOM(trials, probability_s, alpha) returns the smallest $x$ for which the binomial cumulative distribution function (c.d.f.) is greater than or equal to alpha; that is, if $B(x)$ represents the binomial c.d.f. then = CRITBINOM$(n, p, \alpha)$ returns

$$B^{-1}(\alpha) = \inf\{x : B(x) \geq \alpha\}, \quad 0 < \alpha \leq 1.$$

For $\alpha = 0$, this definition gives $-\infty$ and Excel gives the error message #NUM!.

Inverse probabilities are useful for finding $P$-values and in **simulation** because from the definition, if $U$ is uniform $(0, 1)$ and $F(x)$ is an arbitrary c.d.f. with inverse defined by

$$F^{-1}(\alpha) = \inf\{x : F(x) \geq \alpha\}$$

then $X = F^{-1}(U)$ has the specified distribution $F(x)$. Thus, for instance, the Excel formula $= \texttt{CRITBINOM}(u, p, U)$ is a binomial random variable on $n$ trials and success probability $p$, and $= \texttt{NORMINV}(U, \mu, \sigma)$ is a $N(\mu, \sigma)$ random variable.

## Macro – Binomial Simulation

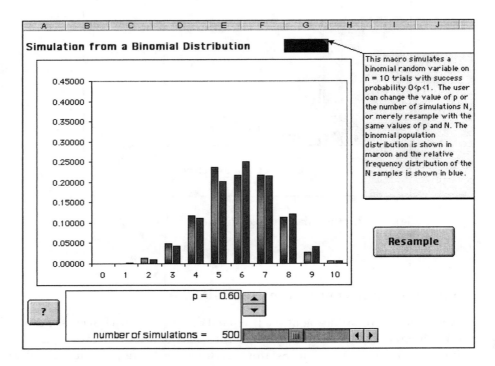

Figure 5.5: Binomial Simulation Macro $n = 10$

The Excel file *bsim.xls* contains a macro that simulates sampling from a binomial random variable on $n = 10$ trials with success probability $0 < p < 1$ and constructs a histogram of the results. The user can change the value of $p$ or the number of simulations $N$ by moving the slider bar and can resample with the same values of $p$ and $N$. This is shown in Figure 5.5, where the binomial probabilities are in the right-most rectangle of each pair and the simulated values are on the left. In the actual files these are color-coded, with the binomial population distribution shown in maroon and the relative frequency distribution of the $N$ samples shown in blue.

## 5.3 The Poisson Distributions

Another important distribution involving counts is the Poisson distribution. It is described by a single parameter $\mu$, the mean. It is a model for events that happen

| | A | B | C | D |
|---|---|---|---|---|
| 1 | Poisson Probabilities | | | |
| 2 | | | | |
| 3 | x | P(X=k) | P(X=k) | P(X<=k) |
| 4 | 0 | =POISSON(A4,1.6,0) | 0.20190 | 0.20190 |
| 5 | 1 | =POISSON(A5,1.6,0) | 0.32303 | 0.52493 |
| 6 | 2 | =POISSON(A6,1.6,0) | 0.25843 | 0.78336 |
| 7 | 3 | =POISSON(A7,1.6,0) | 0.13783 | 0.92119 |
| 8 | 4 | =POISSON(A8,1.6,0) | 0.05513 | 0.97632 |
| 9 | 5 | =POISSON(A9,1.6,0) | 0.01764 | 0.99396 |
| 10 | 6 | =POISSON(A10,1.6,0) | 0.00470 | 0.99866 |
| 11 | 7 | =POISSON(A11,1.6,0) | 0.00108 | 0.99974 |
| 12 | 8 | =POISSON(A12,1.6,0) | 0.00022 | 0.99995 |
| 13 | 9 | =POISSON(A13,1.6,0) | 0.00004 | 0.99999 |
| 14 | 10 | =POISSON(A14,1.6,0) | 0.00001 | 1.00000 |
| 15 | 11 | =POISSON(A15,1.6,0) | 0.00000 | 1.00000 |
| 16 | 12 | =POISSON(A16,1.6,0) | 0.00000 | 1.00000 |
| 17 | | =SUM(C4:C16) | 1.00000 | |

Figure 5.6: Poisson Probabilities $\mu = 1.6$

independently and homogeneously in time or in a region of space. If $X$ is a Poisson distribution with mean $\mu$, then the possible values of $X$ are $0, 1, 2, 3, \ldots$. If $k$ is any non-negative integer, then

$$P(X = k) = \frac{e^{-\mu}\mu^k}{k!} \qquad 0, 1, 2, 3, \ldots$$

The standard deviation of $X$ is $\sqrt{\mu}$.

**Example 5.3.** (Example 5.15, page 346 in text.) A carpet manufacturer knows that the number of flaws per square yard in a type of carpet material varies with an average of 1.6 flaws per square yard. The count $X$ of flaws per square yard can be modeled by the Poisson distribution with $\mu = 1.6$. Calculate the probability that there are no more than two defects in a randomly chosen square yard of this material.

**Solution.** Figure 5.6 shows the relevant formulas and values from an Excel worksheet where the individual and also the cumulative Poisson probabilities are calculated. Such formulas need to be typed only once and then dragged with the mouse cursor to the remaining cells. Excel automatically adjusts the range references, converting A4 into A5, A6, and so on. The formulas, shown here for convenience only, will not be visible on your sheet, only their numerical values will appear. We can read the required answer 0.78336 in cell D6.

# Chapter 6

# Introduction to Inference

One important measure or parameter of a distribution is its mean. This chapter discusses procedures for finding confidence intervals and carrying out significance tests for the mean of a population. The methods require use of the normal distribution and hence are applicable only when the underlying population may be assumed to be approximately normal or when the sample size is so large that the normal approximation given by the central limit theorem may be invoked.

## 6.1  Estimating with Confidence

**Example 6.1.**    (Example 6.2, page 369 in text.)  The ABA survey of community banks asked about the loan-to-deposit ratio, a bank's total loans as a percentage of its total deposits. The mean LTDR for the 110 banks in the sample is $\bar{x} = 76.7$ and the standard deviation is $s = 12.3$. This sample is sufficiently large for us to use $s$ as the population $\sigma$ here. Give a 95% confidence interval for the mean LTDR for community banks.

The bank data $\{x_1, \ldots, x_n\}$ are assumed to come from a $N(\mu, \sigma)$ population with mean $\mu$ and an unknown standard deviation $\sigma$. A level $C$ confidence interval for the mean $\mu$ is given by

$$\bar{x} \pm z^* \frac{\sigma}{\sqrt{n}}$$

where $\bar{x} = \frac{1}{n} \sum_{i=1}^{n} x_i$ is the sample mean, $n$ is the sample size, and $z^*$ is the value such that the area between $-z^*$ and $z^*$ under a standard normal curve equals $C$. In view of the large sample size we are permitted to replace the unknown $\sigma$ with the observed sample standard deviation $s$.

**Solution.**    The Excel function required to compute $z^* \frac{\sigma}{\sqrt{n}}$ is

$$= \mathtt{CONFIDENCE}(\alpha, \sigma, n)$$

which calculates the margin of error (half width of interval) associated with a $C = 1 - \alpha$ level confidence interval for the mean of a normal distribution with standard deviation $\sigma$ based on a sample of size $n$. We enter $= \mathtt{CONFIDENCE}(.05, 12.3, 110)$ into any cell of a worksheet and read off 2.30 which is the margin of error. Then take

$$76.3 \pm 2.30 = (74.40, 79.00)$$

for the 95% confidence interval. It is also possible to calculate each component in the confidence interval directly. This more complicated approach is useful if intermediate results such as the standard error of the estimate are needed. The formula for the normal critical value $z^*$ is $\mathtt{NORMSINV}(p)$, which returns the inverse of the standard normal cumulative distribution function $\Phi^{-1}(p)$, where $0 < p < 1$.

## Macro – Confidence Interval

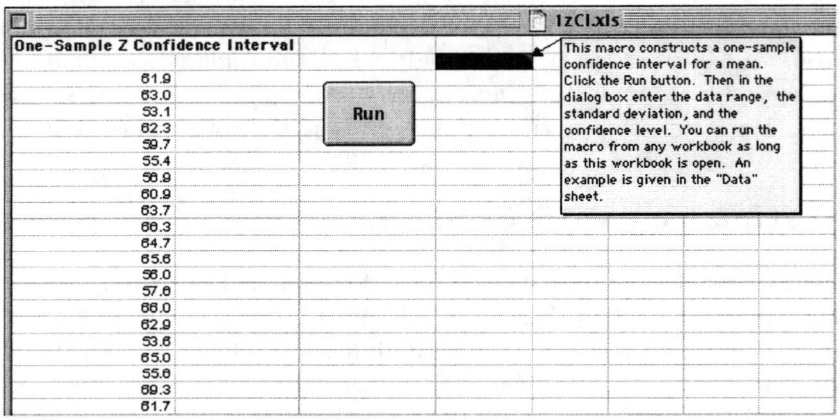

Figure 6.1: Confidence Interval Macro

In Example 6.1 the sample mean was given. With raw data the mean first has to be calculated. To simplify this task we have provided a macro *1zCI.xls* in which the user enters the data and the macro does the computations. We illustrate with an example

**Example 6.2.**    (Example 6.12, page 377 in text.)   Clothing for runners. Your company sells exercise clothing and equipment on the Internet. To design the clothing, you collect data on the physical characteristics of your different types of customers. Here are the weights (in kilograms) for a sample of 24 male runners.

| 67.8 | 61.9 | 63 | 53.1 | 62.3 | 59.7 | 55.4 | 58.9 | 60.9 | 69.2 | 63.7 | 68.3 |
| 64.7 | 65.6 | 56 | 57.8 | 66   | 62.9 | 53.6 | 65   | 55.8 | 60.4 | 69.3 | 61.7 |

Assume these runners can be viewed as a random sample of your potential mail customers. Suppose that the standard deviation of the population is known to be $\sigma = 4.5$ kg. Give a 95% confidence interval for $\mu$, the mean of the population from which the sample is drawn.

**Solution**

1. Open the Excel file *1zCI.xls* and enter the data into a column in that same workbook or in another one. In (Figure 6.1) we have entered the data on the same sheet as the macro.

2. Click the **Run** button to bring up the **One-Sample Z CI** dialog box (Figure 6.2). This macro as well as all others may also be run from the Menu Bar by using **Tools – Macro – Macros** ... and then selecting the appropriate macro from the choices presented, in this case *ZCI*.

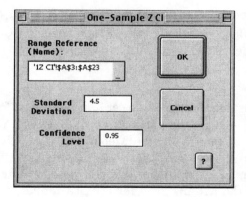

Figure 6.2: Confidence Interval Macro Dialog Box

|  | A | B |
|---|---|---|
| 1 | **User Input** | |
| 2 | standard deviation | 4.5 |
| 3 | confidence level | 0.95 |
| 4 | **Summary Statistics** | |
| 5 | sample size | 21 |
| 6 | sample mean | 61.22 |
| 7 | **Calculations** | |
| 8 | SE | 0.962 |
| 9 | z | 1.960 |
| 10 | ME | 1.925 |
| 11 | | |
| 12 | lower limit | 59.29 |
| 13 | upper limit | 63.14 |

Figure 6.3: Confidence Interval Macro Output

3. Enter the data by selecting the data range, then enter the standard deviation 4.5 and the confidence level 0.95. Click **OK**.

4. The output appears on a new sheet in your workbook (Figure 6.3) from which you can read the answer (59.29, 63.14).

**Note:** Examine the formulas behind the values in the output page to understand the calculation.

## How Confidence Intervals Behave

A confidence interval is a random interval that has a specified probability of containing an unknown parameter. Thus, a 90% confidence interval for a population mean has probability 0.90 of containing the mean. So, in repeated confidence intervals, in the long run approximately 90% of these confidence intervals would contain the population mean.

> **Example 6.3.** Take 100 SRS of size 3 from an $N(3.0, 0.2)$ population and construct a 90% confidence interval for the mean. Count how many times the confidence interval contains the mean 3.0.

**Solution.**

1. Following the instructions given earlier in Example 4.5 for simulating samples from a specified distribution, choose **Tools — Data Analysis — Random Number Generation** from the Menu Bar, complete a box like the one shown in Figure 4.8, but for normal not discrete random numbers, with "3" for the **Number of Variables**, "100" for the **Number of Random Numbers**, "3.0" for the **Mean**, "0.2" for the **Standard Deviation**, and choose a convenient range for the output. We have selected the range A8:C107.

| | A | B | C | D | E | F | G | H |
|---|---|---|---|---|---|---|---|---|
| 1 | | **Behavior of Repeated Confidence Intervals** | | | | | | |
| 2 | | | | | | | | |
| 3 | lower | = AVERAGE(A8:C8) −NORMSINV(0.5+0.90/2)*0.2/SQRT(3) | | | | | | |
| 4 | upper | = AVERAGE(A8:C8) −NORMSINV(0.5+0.9/2)*0.2/SQRT(3) | | | | | | |
| 5 | | G8 =IF(AND(E8<3, 3<F8), 1,0) | | | | | | |
| 6 | | | | | | | | |
| 7 | | | | | lower | upper | | |
| 8 | 3.1772 | 2.7218 | 3.3097 | | 2.880 | 3.259 | 1 | 92 |
| 9 | 3.0863 | 3.0417 | 2.8220 | | 2.793 | 3.173 | 1 | |
| 10 | 2.8207 | 2.8480 | 2.9353 | | 2.678 | 3.058 | 1 | |
| 11 | 2.9131 | 3.2380 | 3.0292 | | 2.870 | 3.250 | 1 | |
| 12 | 3.0904 | 3.1497 | 2.9295 | | 2.867 | 3.246 | 1 | |
| 13 | 2.8767 | 3.0868 | 3.2555 | | 2.883 | 3.263 | 1 | |
| 14 | 2.8937 | 2.7254 | 2.9995 | | 2.683 | 3.063 | 1 | |
| 15 | 3.1976 | 3.0303 | 2.8750 | | 2.844 | 3.224 | 1 | |
| 16 | 2.8378 | 2.9206 | 2.7565 | | 2.648 | 3.028 | 1 | |
| 17 | 2.7972 | 2.9133 | 3.1956 | | 2.779 | 3.159 | 1 | |
| 18 | 2.5773 | 3.2215 | 3.0810 | | 2.770 | 3.150 | 1 | |
| 19 | 3.1855 | 2.6901 | 3.0221 | | 2.776 | 3.156 | 1 | |
| 20 | 3.1730 | 3.1092 | 3.0020 | | 2.905 | 3.285 | 1 | |

Figure 6.4: Repeated Confidence Intervals

2. In cell E8 enter
   $= \texttt{AVERAGE(A8:C8)} - \texttt{NORMSINV}(0.5+0.9/2)\texttt{*0.2/SQRT(3)}$
   In cell F8 enter
   $= \texttt{AVERAGE(A8:C8)} + \texttt{NORMSINV}(0.5+0.9/2)\texttt{*0.2/SQRT(3)}$

3. Select cells E8:F8, click the fill handle, and drag the contents to F107. The cells in column F will contain the value 1 if the confidence interval for the data in the corresponding row contains the true value 3.0, otherwise the cells will contain 0.

4. Count the number of times 1 appears by entering $= \texttt{SUM(G8:G107)}$ in an empty cell (H8, for example).

Figure 6.4 shows a portion of a workbook with the simulation for which 92 times out of 100, the true mean was within the 90% confidence limits.

## 6.2  Tests of Significance

While a confidence interval is the appropriate technique to use when the objective is to estimate a parameter, a significance test is the tool for judging whether a specified (null) hypothesis is consistent with a data set. A significance test compares the observations with what would be expected if the null hypothesis were true and therefore judges how well the hypothesis fits with the data. If the fit is not good then the hypothis is thrown out (rejected).

## Macro – One-Sample $Z$ Test

We have created a macro *1zT.xls* for carrying out a hypothesis test. It is used in exactly the same fashion as the previous one for constructing confidence intervals. The data may be entered on a worksheet of the macro workbook or in another workbook. The user clicks the **Run** button, completes the dialog box, and clicks **OK**. The output appears in a new sheet of the macro workbook.

The data $\{x_1, x_2, \ldots, x_n\}$ are assumed to come from an $N(\mu, \sigma)$ population where $\sigma$ is known. The same procedure can also be used to carry out a large sample test. The user inputs are the null value, the standard deviation, and the level of significance. The output also includes the $P$-value.

> **Example 6.4.** (Example 6.17, page 397 in text.) Bottles of a popular cola drink are supposed to contain 300 ml of cola. There is some variation from bottle to bottle because the filling machinery is not perfectly precise. The distribution of the contents is normal with standard deviation $\sigma = 3$ ml. A student who suspects that the bottle is underfilling measures the contents of six bottles. The results are
>
> $$299.4 \quad 297.7 \quad 310.0 \quad 298.9 \quad 300.2 \quad 297.0.$$

Is this convincing evidence that the mean content of cola bottles is less than the advertised 300 ml?

**Solution.**     The null hypothesis is $H_0 : \mu = 300$ and the alternative is $H_a : \mu < 300$.

1. Enter the six data values into a worksheet.

2. Click the **Run** button for the macro to bring up the dialog box (Figure 6.5) and complete it as shown by selecting the data range and entering the null value, the standard deviation, and the level of significance "alpha". Then click **OK**.

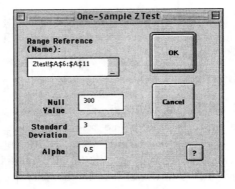

Figure 6.5: Significance Test Macro Dialog Box

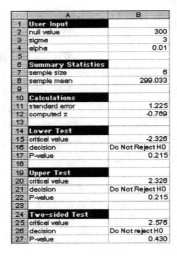

Figure 6.6: Significance Test Macro Dialog Box

3. The output (Figure 6.6) shows three regions depending on the type of alternative. Ours is a lower tail test so the relevant portion gives a 1% critical value of -2.326, a computed $z$ value of 0.789 (which differs from the value 0.792 in the text because the latter was obtained by rounding the sample mean to 299.03), and a $P$-value of 0.215. The conclusion is "Do not reject" $H_0$.

## Explanation

The Excel output from the macro shows all formulas required in column B for any type of alternative: lower, upper, and two-sided tests, respectively. It is instructive for the user to examine these. The function NORMSINV was encountered previously. The formula = NORMSDIST returns the cumulative normal distribution function $\Phi(z)$. For a one-sided lower test, the $P$-value is the area to the left of the computed $z$ score $\frac{\bar{x}-\mu_0}{\sigma/\sqrt{n}}$ and is thus given by = NORMSDIST$(z)$. For an upper test, the $P$-value is the area to the right of $\frac{\bar{x}-\mu_0}{\sigma/\sqrt{n}}$ and is given by = $1 -$ NORMSDIST$(z)$. For a two-sided test, the formula for the $P$-value is = 2*(1 $-$ NORMSDIST$(|z|)$). The formula = ABS$(z)$ returns the absolute value of $z$. The decision rule uses the logical = IF(statement, true, false), which returns the string designated as true if the statement is true or else it returns false.

# Chapter 7

# Inference for Distributions

In this chapter we continue to develop inference procedures for the mean of a population. While in Chapter 6 we assumed that the population standard deviation $\sigma$ was known, here we do not make this demand. Instead we estimate $\sigma$ through the sample standard deviation $s$. For normal populations, inference based on the standard normal $Z$ is replaced with inference based on Student's $t$ distribution. For instance, a level $C$ confidence interval for the mean becomes

$$\bar{x} \pm t^* \frac{s}{\sqrt{n}}$$

while a significance test is based on the "signal-to-noise" ratio

$$t = \frac{\bar{x} - \mu}{s/\sqrt{n}}$$

As before $n, \bar{x}$ are the sample size and sample mean respectively, and $t^*$ is the critical $t$ value such that the area between $-t^*$ and $t^*$ under the curve of a $t$ density with $n-1$ degrees of freedom equals $C$. The Excel formula required to calculate critical values is

$$= \text{TINV}(\alpha, \nu)$$

which returns the critical value for a level $C = 1 - \alpha$ confidence interval based on a $t$ distribution with $\nu$ degrees of freedom.

We also develop procedures for inference on the difference in two population means that will require use of the **Analysis ToolPak**.

## 7.1  Inference for the Mean of a Population

Our first example shows how to find a confidence interval for a mean. It uses a macro developed for this purpose.

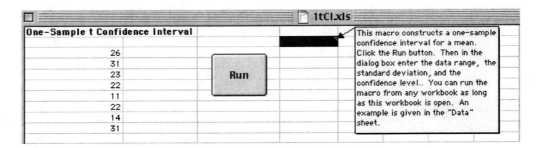

Figure 7.1: Confidence Interval Macro

# Macro – One-Sample $t$ Confidence Interval

**Example 7.1.** (Case 7.1 and Example 7.1, pages 435–436 in text.) Many food products are fortified by adding nutrients, especially vitamins. In a recent year, the U.S. Agency for International Development purchased 238,300 metric tons of corn soy blend (CSB) for development programs and emergency relief in countries throughout the world. CSB is a highly nutritious, low cost fortified food that is partially pre-cooked and can be incorporated into different food preparations by the recipient. The addition of the nutrients is a difficult and important part of the production process. If too little is present, the product will be inefffective. If too much is present, consumption of the product could have some adverse effects. Vitamin C is an important nutrient used to fortify CSB. A study of the quality of the production process of CSB therefore measured vitamin C in specimens taken at the factory. The data given below are the amounts of vitamin C, in milligrams per 100 grams of dry blend for a random sample of size 8 from a production run.

26   31   23   22   11   22   14   31

Find a 95% confidence interval for $\mu$, the mean vitamin C content of the CSB produced during the run.

**Solution.** The details parallel those described earlier for Example 6.2 when $\sigma$ was known.

1. Open the Excel macro file *1tCI.xls* and enter the data into a column in that same workbook or in another one. In (Figure 7.1) we have entered the data on the same sheet as the macro.

2. Click the **Run** button to bring up the **One-Sample t** confidence interval dialog box (Figure 7.2). This macro as well as all others may also be run

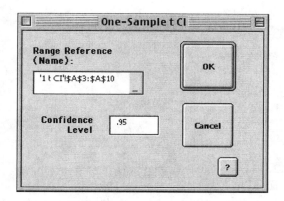

Figure 7.2: Confidence Interval Macro Dialog Box

from the Menu Bar using **Tools − Macro − Macros** ... and then selecting
the appropriate macro from the choices presented.

|    | A | B |
|----|---|---|
| 1 | **User Input** | |
| 2 | confidence level | 0.95 |
| 3 | **Summary Statistics** | |
| 4 | sample size | 8 |
| 5 | sample mean | 22.5 |
| 6 | sample standard deviation | 7.191 |
| 7 | **Calculations** | |
| 8 | SE | 2.542 |
| 9 | df | 7 |
| 10 | t | 2.365 |
| 11 | ME | 6.012 |
| 12 | **Confidence Limits** | |
| 13 | lower limit | 16.488 |
| 14 | upper limit | 28.512 |

Figure 7.3: Confidence Interval Macro Output

3. Enter the data by selecting the data range, then enter the standard deviation
   4.5 and the confidence level 0.95. Click **OK**.

4. The output appears on a new sheet in your workbook (Figure 7.3) from which
   you can read the answer (16.48, 28.51).

**Note:** Examine the formulas behind the values in the output page to understand
the calculation. We have calculated the sample standard deviation $s$ using the
Excel formula $= \texttt{STDEV}(\text{Data})$. The critical value $t^*$ (denoted by $t$ on the workbook)
is obtained from the formula $= \texttt{TINV}(1 - conf, df)$, where $conf$ is the confidence
level and $df = n - 1$ are the degrees of freedom.

# Macro – One-Sample $t$ Test

For the significance test

$$H_0 : \mu = \mu_0$$

the test statistic is

$$t = \frac{\bar{x} - \mu_0}{s/\sqrt{n}}$$

which has Student's $t$ distribution on $n - 1$ degrees of freedom.

> **Example 7.2.** (Example 7.2, page 437 in text.) The specifications for the CSB described in Example 7.1 state that the mixture should contain two pounds of vitamin premix for every 2000 pounds of product. These specifications are designed to produce a mean ($\mu$) vitamin C content in the final product of 40 mg/100 g. We test the null hypothesis that the mean vitamin C content is 40 mg/100 g.
>
> $$\begin{aligned} H_0 : \mu &= 40 \\ H_a : \mu &\neq 40 \end{aligned}$$

**Solution.**

1. Open the macro file *1tT.xls* and enter the eight data values into a worksheet.

2. Click the **Run** button for the macro to bring up the dialog box (Figure 7.4 and complete it as shown by selecting the data range and entering the null value and the level of significance "alpha." Then click **OK**.

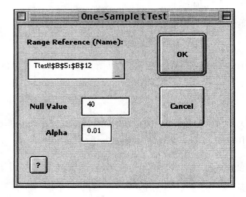

Figure 7.4: Significance Test Macro Dialog Box

3. The output shows three regions, depending on the type of alternative. Ours is a two-tailed test so the relevant portion gives a 1% critical value of -3.499, a computed $t$ value of -6.883, and a $P$-value of 0.0002. The conclusion is "Reject" $H_0$.

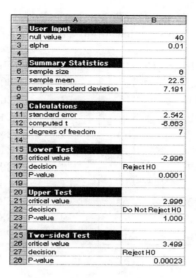

| | A | B |
|---|---|---|
| 1 | **User Input** | |
| 2 | null value | 40 |
| 3 | alpha | 0.01 |
| 4 | | |
| 5 | **Summary Statistics** | |
| 6 | sample size | 8 |
| 7 | sample mean | 22.5 |
| 8 | sample standard deviation | 7.191 |
| 9 | | |
| 10 | **Calculations** | |
| 11 | standard error | 2.542 |
| 12 | computed t | -6.883 |
| 13 | degrees of freedom | 7 |
| 14 | | |
| 15 | **Lower Test** | |
| 16 | critical value | -2.998 |
| 17 | decision | Reject H0 |
| 18 | P-value | 0.0001 |
| 19 | | |
| 20 | **Upper Test** | |
| 21 | critical value | 2.998 |
| 22 | decision | Do Not Reject H0 |
| 23 | P-value | 1.000 |
| 24 | | |
| 25 | **Two-sided Test** | |
| 26 | critical value | 3.499 |
| 27 | decision | Reject H0 |
| 28 | P-value | 0.00023 |

Figure 7.5: Significance Test Macro Dialog Box

**Note:** Excel has an *unusual* definition for the cumulative $t$, namely,

$$\text{TDIST}(x, \nu, 1) = P[t(\nu) > x]$$

where $t(\nu)$ is a $t$ distribution on $\nu$ degrees of freedom. The argument $x$ must be positive, and this accounts for the more complicated syntax in the decision rule in the corresponding template. See the formulas in the output from the macro.

## Matched Pairs $t$ Procedure

In order to reduce variability in a data set, scientists sometimes use paired data matched on characteristics believed to affect the response. This is equivalent to a randomized block design. Such data are best analyzed if one-sample procedures are applied to differences between the pairs. There is a loss in degrees of freedom for error, but if the matching is effective, this will be more than offset by the gain in reduced variance of the differences. The same method applies to before-after measurements on the same subjects.

> **Example 7.3.** (Example 7.7, page 443 in text.) A company contracts with a language institute to provide individualized instruction in foreign languages for its executives who will be posted overseas. Is instruction effective? Last year, 20 executives studied French. All had some knowledge of French, so they were given Modern Languages Association's listening test of understanding of spoken French before

Table 7.1: French listening scores for executives

| Teacher | Pretest | Posttest | Teacher | Pretest | Posttest |
|---------|---------|----------|---------|---------|----------|
| 1 | 32 | 34 | 11 | 30 | 36 |
| 2 | 31 | 31 | 12 | 20 | 26 |
| 3 | 29 | 35 | 13 | 24 | 27 |
| 4 | 10 | 16 | 14 | 24 | 24 |
| 5 | 30 | 33 | 15 | 31 | 32 |
| 6 | 33 | 36 | 16 | 30 | 31 |
| 7 | 22 | 24 | 17 | 15 | 15 |
| 8 | 25 | 28 | 18 | 32 | 34 |
| 9 | 32 | 26 | 19 | 23 | 26 |
| 10 | 20 | 26 | 20 | 23 | 26 |

instruction began. After several weeks of immersion in French, the executives took the listening test again. The pretest and posttest scores are provided in Table 7.1. Let $\mu$ denote the mean improvement that would be achieved if the entire population of executives received similar instruction. We wish to test at the 0.01 level of significance

$$H_0 : \mu = 0$$
$$H_a : \mu > 0$$

## Matched Pairs *t* Test—Using the ToolPak

To analyze these data, subtract the pretest scores from the posttest scores for each teacher. The 20 differences form a single sample. Thus, we may apply the previous one-sample macros to these differences. Excel also provides a direct method for the matched pairs *t* test using the **Analysis ToolPak**, which we now describe. However, *it cannot be used with summarized data.*

1. Open a new workbook and enter the Pretest and Posttest values with their labels in cells A3:A23 and B3:B23, respectively, as in Figure 7.6.

2. In cell C3 enter the label "Difference." In cell C4 type the formula $= A4 - B4$. Then select C4 and fill to cells C4:C23.

3. Choose **Tools – Data Analysis** from the Menu Bar, and then check the box *t*-**Test:Paired Two Sample for Means**. Click **OK**.

4. Complete the next dialog box (Figure 7.7). **Caution** is required in determining which scores are entered for **Variable 1 range** and which scores are entered for **Variable 2 range**. Excel takes Variable 1 - Variable 2 as the difference.

Because we are taking posttest − pretest scores, enter the posttest score range B3:B23 for Variable 1 and the pretest range A3:A23 for Variable 2.

## Excel Output

| | A | B | C | D | E | F |
|---|---|---|---|---|---|---|
| 1 | | | T Test for a Normal Mean (Using the Analysis ToolPak) | | | |
| 2 | | | | | | |
| 3 | Pretest | Posttest | Difference | t-Test: Paired Two Sample for Means | | |
| 4 | 32 | 34 | 2 | | | |
| 5 | 31 | 31 | 0 | | Posttest | Pretest |
| 6 | 29 | 35 | 6 | Mean | 28.3 | 25.8 |
| 7 | 10 | 16 | 6 | Variance | 35.379 | 39.747 |
| 8 | 30 | 33 | 3 | Observations | 20 | 20 |
| 9 | 33 | 36 | 3 | Pearson Correlation | 0.890 | |
| 10 | 22 | 24 | 2 | Hypothesized Mean Difference | 0 | |
| 11 | 25 | 28 | 3 | df | 19 | |
| 12 | 32 | 26 | -6 | t Stat | 3.865 | |
| 13 | 20 | 26 | 6 | P(T<=t) one-tail | 0.00052 | |
| 14 | 30 | 36 | 6 | t Critical one-tail | 2.539 | |
| 15 | 20 | 26 | 6 | P(T<=t) two-tail | 0.00104 | |
| 16 | 24 | 27 | 3 | t Critical two-tail | 2.861 | |
| 17 | 24 | 24 | 0 | | | |
| 18 | 31 | 32 | 1 | | | |
| 19 | 30 | 31 | 1 | | | |
| 20 | 15 | 15 | 0 | | | |
| 21 | 32 | 34 | 2 | | | |
| 22 | 23 | 26 | 3 | | | |

Figure 7.6: Paired *t* Test—Analysis ToolPak

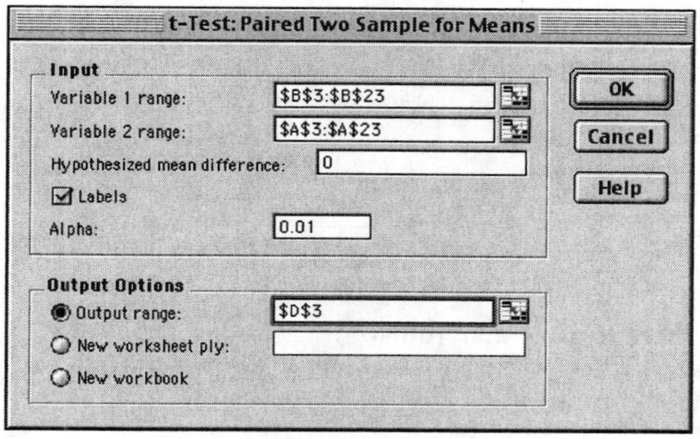

Figure 7.7: Paired *t* Test—Dialog Box

The output appears in Figure 7.6 in the range D3:F16 beginning with cell D3, as specified in Figure 7.7. Individual sample means are given as well as the test statistic value $t$ in cell E10. One-sided and two-sided critical values are provided in E14 and E16, as well as corresponding $P$-values in E13 and E15. Because $H_0 : \mu > 0$ and $t = 3.865$, we reject at level $\alpha = 0.01$. The $P$-value is 0.00052.

The entry $P(T <= t)$ one-tail in E13 is not $P(t(19) \leq 3.865)$, as the notation would suggest. Rather $P(T <= t)$ represents the tail area relative to $t$ (so it is a lower tail if $t$ is negative and an upper tail if $t$ is positive). It is therefore not always a $P$-value.

### Confidence Interval for Paired Data

The **Analysis ToolPak** does not provide a confidence interval directly but the one sample $t$ confidence interval macro *1tCI.xls* can do this, using the differences between posttest and pretest scores.

> **Exercise.** For Example 7.3, show that a 90% confidence interval for the difference in posttest mean and pretest mean is (1.382, 3.618).

## 7.2  Comparing Two Means

Independent simple random samples of sizes $n_1$ and $n_2$ are obtained from populations with means $\mu_1$ and $\mu_2$, respectively. We are interested in comparing $\mu_1$ and $\mu_2$. The appropriate statistics and critical values required depend on the assumptions made. Suppose that the data are collected from normal populations with standard deviations $\sigma_1$ and $\sigma_2$. Denote by $\bar{x}_1, s_1^2, \bar{x}_2$, and $s_2^2$ the corresponding summary statistics (the sample means and sample variances).

Inference is usually based on a two-sample statistic of the form

$$\frac{(\bar{x}_1 - \bar{x}_2) - (\mu_1 - \mu_2)}{\text{SE}} \tag{7.1}$$

where SE represents the standard error of the numerator.

If the underlying populations are normal with known population standard deviations $\sigma_1$ and $\sigma_2$, then we use

$$z = \frac{(\bar{x}_1 - \bar{x}_2) - (\mu_1 - \mu_2)}{\sqrt{\frac{\sigma_1^2}{n_1} + \frac{\sigma_2^2}{n_2}}}$$

which has a standard normal distribution.

If, however, $\sigma_1$ and $\sigma_2$ are unknown, then the appropriate ratio is

$$t = \frac{(\bar{x}_1 - \bar{x}_2) - (\mu_1 - \mu_2)}{\sqrt{\frac{s_1^2}{n_1} + \frac{s_2^2}{n_2}}} \tag{7.2}$$

which is called a two-sample $t$ statistic.

Actually, the distribution of the statistic in (7.2) depends on $\sigma_1$ and $\sigma_2$ and does not have an exact $t$ distribution. Nonetheless, it is used with $t$ critical values in inference in one of two ways, each involving a computed value for degrees of freedom $\nu$ associated with the denominator of (7.1) to provide an approximate $t$ statistic. The two options follow.

1. Use a value for $\nu$ given by

$$\nu = \frac{\left(\frac{s_1^2}{n_1} + \frac{s_2^2}{n_2}\right)^2}{\frac{1}{n_1-1}\left(\frac{s_1^2}{n_1}\right)^2 + \frac{1}{n_2-1}\left(\frac{s_2^2}{n_2}\right)^2} \tag{7.3}$$

2. Use

$$\nu = \min\{n_1 - 1, n_2 - 1\} \tag{7.4}$$

Approximation (7.4) is conservative in the sense of providing a larger margin of error and is recommended in the text when doing calculations without the aid of software. Approximation (7.3) is considered to provide a quite accurate approximation to the actual distribution.

Excel, like most statistical software, uses the value for $\nu$ given in (7.3) and provides three tools in the **Analysis ToolPak** for analyzing independent samples from normal populations: $z$ test, $t$ test (unequal variances), and pooled $t$ test (equal variances). These involve different sets of assumptions and consequently different forms for the SE. Each tool provides a similar dialog box for user input, parameters, and the range for the actual raw data.

However, sometimes only summary statistics are available and the **Analysis ToolPak** cannot be used; instead, direct calculations are required.

## Two-Sample $z$ Statistic

For normal populations with known standard deviations, the "standard score"

$$z = \frac{(\bar{x}_1 - \bar{x}_2) - (\mu_1 - \mu_2)}{\sqrt{\frac{\sigma_1^2}{n_1} + \frac{\sigma_2^2}{n_2}}}$$

has a standard normal distribution. This ratio is also used for "large sample" procedures with $\sigma_1$ and $\sigma_2$ unknown and replaced by the sample standard deviations $s_1$ and $s_2$. Normality of the underlying population is not mandatory for large sample tests, and the corresponding confidence intervals and significance tests have approximately the specified values.

### Two-Sample $z$ Confidence Interval

**Example 7.4.** (Exercise 7.76(b), page 484 in text.) Does cocaine use by pregnant women cause their babies to have low birth weight? To study this question, birth weights of babies of women who tested positive for cocaine/crack during a drug-screening test were compared with the birth weights for women who either tested negative or were not tested, a group we call "other." Here are the summary statistics. The birth weights are measured in grams.

| Group | $n$ | $\bar{x}$ | $s$ |
|---|---|---|---|
| Positive test | 134 | 2733 | 599 |
| Others | 5974 | 3118 | 672 |

Give a 95% confidence interval for the mean difference in birth weights.

| | A | B | C | D |
|---|---|---|---|---|
| 1 | **Two-Sample Z Confidence Interval** | | | |
| 2 | | | | |
| 3 | | Summary Statistics and User Input | | |
| 4 | Group | n | xbar | s |
| 5 | Positive | 134 | 2733 | 599 |
| 6 | Other | 5974 | 3118 | 672 |
| 7 | | | | |
| 8 | conf | 0.95 | | |
| 9 | SE | 52.471 | =SQRT(D5^2/B5+D6^2/B6) | |
| 10 | z | 1.960 | =NORMSINV(0.5+conf/2) | |
| 11 | ME | 102.841 | =z*SE | |
| 12 | Confidence Limits | | | |
| 13 | lower | -487.84 | =(C5-C6)-ME | |
| 14 | upper | -282.16 | =(C5-C6)+ME | |

Figure 7.8: Large Two-Sample $z$ Confidence Interval

**Solution.** We will use a procedure based on a large sample $z$, treating the observed sample standard deviations as the corresponding population values. Figure 7.8 provides both formulas (column C) to derive the required confidence interval and the values obtained (column B). The user inputs are $\bar{x}_1$, $\bar{x}_2$, $\sigma_1$, $\sigma_2$, $n_1$, $n_2$, and the confidence level $C$. We have **Named** the ranges to refer to "conf," "SE," "z," and "ME." We can read off from cells B13:B14 that a 95% confidence interval is $(-487.84, -282.16)$.

## Two-Sample $z$ Test

> **Example 7.5.** (Continuing Exercise 7.76, page 484 in text.) Test the hypothesis that babies of women who use cocaine while pregnant have a lower birth weight on average than babies of women in the "other" group.

**Solution.** The hypothesis to be tested is

$$H_0 : \mu_1 = \mu_2$$
$$H_a : \mu_1 < \mu_2$$

where $\mu_1$ and $\mu_2$ are the hypothesized mean birth weights for babies born to women who test positive and the "others," respectively. At the level $\alpha = 0.05$, the decision rule is to reject $H_0$ if the computed $z$ value

$$z = \frac{\bar{x}_1 - \bar{x}_2}{\sqrt{\dfrac{\sigma_1^2}{n_1} + \dfrac{\sigma_2^2}{n_2}}}$$

satisfies $z < 1.645$. The $P$-value will be given by

$$P\text{-value } = \Phi(z)$$

where $\Phi(z)$ is the cumulative $N(0,1)$ distribution function. As in the previous example, in view of the large sample sizes, we can use $s_1$ and $s_2$, the sample standard deviations, in place of $\sigma_1$ and $\sigma_2$. These calculations can readily be made by a hand calculator.

| | A | B | C | D | F |
|---|---|---|---|---|---|
| 1 | | **Large Two-Sample Z Test** | | | |
| 2 | | | | | |
| 3 | | Summary Statistics and User Input | | | |
| 4 | Group | n | xbar | s | |
| 5 | Positive | 134 | 2733 | 599 | |
| 6 | Other | 5974 | 3118 | 672 | |
| 7 | | | | | |
| 8 | SE | 52.471 | =SQRT(D5^2/B5+D6^2/B6) | | |
| 9 | z | -7.337 | =((C5-C6)-null)/SE | | |
| 10 | alpha | 0.05 | | | |
| 11 | null | 0 | | | |
| 12 | Lower Test | | | | |
| 13 | lower_z | -1.645 | =NORMSINV(alpha) | | |
| 14 | Decision | Reject H0 | =IF(z<lower_z,"Reject H0","Do Not Reject H0") | | |
| 15 | Pvalue | 1.097E-13 | =NORMSDIST(z) | | |
| 16 | Upper Test | | | | |
| 17 | upper_z | | =-NORMSINV(alpha) | | |
| 18 | Decision | | =IF(z>upper_z,"Reject H0","Do Not Reject H0") | | |
| 19 | Pvalue | | =1-NORMSDIST(z) | | |
| 20 | Two-Sided Test | | | | |
| 21 | two_z | | =ABS(NORMSINV(alpha/2)) | | |
| 22 | Decision | | =IF(ABS(z)>two_z,"Reject H0","Do Not Reject H0") | | |
| 23 | Pvalue | | =2*(1-NORMSDIST(ABS(z))) | | |

Figure 7.9: Large Two-Sample $z$ Test

While the full power of Excel comes from dealing with large data sets, even this simple example can illustrate the use of a spreadsheet to evaluate formulas within equations. Figure 7.9 is an Excel workbook containing a template for this type of problem, showing the Excel formulas required. The calculations are carried out in column B and the corresponding formulas are displayed in the adjacent cells in column C. The problem at hand is a lower test, but formulas are additionally provided for both upper and two-sided tests, only one of which should be used at any time. For purposes of clarity, we have used **Named Ranges** rather than cell references to refer to $\alpha = $ alpha in cell B10, to $z$ in B9, to the standard error

$$SE = \sqrt{\frac{s_1^2}{n_1} + \frac{s_2^2}{n_2}}$$

in B8, and to the critical values lower_z, upper_z, and two_z for the three types of alternative hypotheses. Remember to name your ranges when copying and adapting this template to your own workbook.

The Excel output appears in Figure 7.9 in cells B8:B15 and the lower 5% critical value of -1.645 is in B13. The conclusion "Reject $H_0$" is printed by Excel in B14, and the $P$-value, $1.097 \times 10^{-13}$, is also evaluated in B15.

### Two-Sample $z$ Inference Using the ToolPak

While the aforementioned templates have been designed for summarized data, they can easily be modified for use with raw data, for instance, by entering = AVERAGE(Range) in place of the *value* of the sample mean, where Range is the cell range for the sample. Because population standard deviations are seldom known but instead are replaced by their sample estimates in most applications, if you are dealing with large raw data sets whose standard deviations have not yet been calculated, it is simplest to let Excel then do the calculation. Enter = STDEV(Range) where you would enter the value for the standard deviation and proceed as before. Excel computes the standard deviation from the data and inserts it where required in the formulas.

There is a two-sample $z$ test option included in the **Analysis ToolPak**. But for large sample sizes, the results of two-sample $t$ and $z$ options are virtually the same, and the sequence of steps in the two tools are identical. Therefore, as we will discuss the two-sample $t$ ToolPak in detail in the next section, we will not illustrate its $z$ counterpart. Also, there is a bug in the two-sample $z$ ToolPak that outputs an incorrect two-sided $P$-value; moreover, there is no built-in confidence interval procedure. These considerations limit the usefulness of the two-sample $z$ ToolPak, and we do not recommend using it.

## Two-Sample $t$ Procedures

### Using the Analysis ToolPak

Excel provides three tools in the **Analysis ToolPak** for comparing means from two populations, based on independent samples. These are the two-sample $t$ test, pooled two-sample $t$ test, and two-sample $z$ test. These tools provide dialog boxes in which the user locates the data and decides on the type of analysis desired.

> **Example 7.6.** (Example 7.11, page 464 in text.) A company that sells educational materials reports statistical studies to convince customers that its materials will improve learning. One new product supplies "directed reading activities" for classroom use. These activities should improve the reading ability of elementary school pupils. A consultant arranges for a third-grade class of 21 students to take part in these activities for an eight-week period. A control classroom of 23 third graders follows the same curriculum with the activities. At the end of eight weeks, all students are given a Degree of Reading Power (DRP) test, which measures the aspects of reading ability that the

Table 7.2: DRP scores for third-graders

| Treatment group | | | | Control group | | | |
|---|---|---|---|---|---|---|---|
| 24 | 61 | 59 | 46 | 42 | 33 | 46 | 37 |
| 43 | 44 | 52 | 43 | 43 | 41 | 10 | 42 |
| 58 | 67 | 62 | 57 | 55 | 19 | 17 | 55 |
| 71 | 49 | 54 | | 26 | 54 | 60 | 28 |
| 43 | 53 | 57 | | 62 | 20 | 53 | 48 |
| 49 | 56 | 33 | | 37 | 85 | 42 | |

treatment is designed to improve. The data appear in Table 7.2. Test the hypothesis that directed reading improves scores.

**Solution.**

1. Open a new workbook and enter the data from Table 7.3 on page 530 of the text. Insert the treatment group in cells A4:A24 and the control group in Cells B4:B26. Enter the label "Treatment" in A3 and the label "Control" in B3. (Refer to Figure 7.12, which also shows the output.)

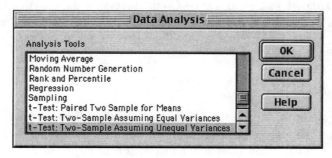

Figure 7.10: Two-Sample $t$ Test Analysis Tools

2. From the Menu Bar choose **Tools − Data Analysis** and select $t$-**Test: Two-Sample Assuming Unequal Variances** from the list of selections (Figure 7.10). Click **OK** (equivalently, double-click your selection).

3. A dialog box (Figure 7.11) appears. Complete as shown. **Variable 1 range** refers to the cell addresses of the sample you have designated by subscript 1, in this case the treatment group. Type A3:A24 in its text area (with the flashing vertical I-beam). Alternatively, you can point to the data by clicking on cell A3 and dragging to the end of the treatment data, cell A24. The values $A$3:$A$24 will appear in the text area of the dialog box. Similarly, enter the range B3:B26 for the control group in the **Variable 2 range**.

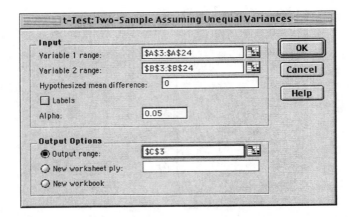

Figure 7.11: Two-Sample $t$ Test Dialog Box

4. The **Hypothesized mean difference** refers to the null value, which is 0 here. Check the **Labels** box, because your ranges included the labels for the two groups. The level of significance **Alpha** is the default 0.05. We will be placing the output in the same workbook as the data, so check the radio button **Output range** and type C3 in the text area. Finally, click **OK**. The output will appear in a block of cells whose upper left corner is cell C3 in Figure 7.12.

## Excel Output

The output appears in the range C3:E15 in Figure 7.12. The range C18:E23 is not part of the output but is the result of additional formulas we have entered to give confidence intervals (discussed next). From cells D6:E6, we see that the sample means for treatment and control groups are 51.476 and 41.522, while the sample variances are 121.162 and 294.079, respectively. The degrees of freedom are 38 (Excel rounds up to the nearest integer). The computed $t$ statistic in cell D11 is 2.311, while the one-sided critical $t^*$ value on 38 degrees of freedom at the 5% level in cell D13 is 1.686. Because the computed $t$ exceeds $t^*$, we reject the null hypothesis and conclude that there is strong evidence that the directed reading activities help elementary school pupils improve some aspects of their reading ability.

Excel provides $P$-values in cells D12 and D14. Here the $P$-value is 0.0132. As before, the entry "$P(T <= t)$ one-tail" in C12 needs some explanation. It is meant to be a one-tailed $P$-value, which depends on the calculated $t$ Stat, the computed $t$ statistic. If $t$ Stat $< 0$, then $P(T <= t)$ one-tail is in fact the lower tail corresponding to the area to the left of $t$ Stat under a $t$ density curve. But if $t$ is positive, then $P(T <= t)$ one-tail is the area to the right of $t$ Stat. It is therefore

| | A | B | C | D | E | F |
|---|---|---|---|---|---|---|
| 1 | | | **Two-Sample t Test** | | | |
| 2 | | | | | | |
| 3 | Treatment | Control | t-Test: Two-Sample Assuming Unequal Variances | | | |
| 4 | 24 | 10 | | | | |
| 5 | 33 | 17 | | *Treatment* | *Control* | |
| 6 | 43 | 19 | Mean | 51.476 | 41.522 | |
| 7 | 43 | 20 | Variance | 121.162 | 294.079 | |
| 8 | 43 | 26 | Observations | 21 | 23 | |
| 9 | 44 | 28 | Hypothesized Mean Difference | 0 | | |
| 10 | 46 | 33 | df | 38 | | |
| 11 | 49 | 37 | t Stat | 2.311 | | |
| 12 | 49 | 37 | P(T<=t) one-tail | 0.013 | | |
| 13 | 52 | 41 | t Critical one-tail | 1.686 | | |
| 14 | 53 | 42 | P(T<=t) two-tail | 0.026 | | |
| 15 | 54 | 42 | t Critical two-tail | 2.024 | | |
| 16 | 56 | 42 | | | | |
| 17 | 57 | 43 | | | | |
| 18 | 57 | 46 | Mean Difference | 9.954 | =D6-E6 | |
| 19 | 58 | 48 | SE | 4.308 | =SQRT(D7/D8+E7/E8) | |
| 20 | 59 | 53 | t | 2.024 | =TINV(0.05,D10) | |
| 21 | 61 | 54 | ME | 8.72 | =D20*D19 | |
| 22 | 62 | 55 | lower | 1.23 | =D18-D21 | |
| 23 | 67 | 55 | upper | 18.67 | =D18+D21 | |
| 24 | 71 | 60 | | | | |
| 25 | | 62 | | | | |
| 26 | | 85 | | | | |

Figure 7.12: Two-Sample $t$ Test ToolPak Output

not always the $P$-value. For instance, if the test to be carried out were

$$H_0 : \mu_1 = \mu_2$$
$$H_a : \mu_1 < \mu_2$$

then the $P$-value would be $1 - 0.0132 = 0.9868$ rather than $0.0121$. $P(T <= t)$ two-tail is the correct $P$-value for a two-tailed test.

## Confidence Intervals

**Example 7.7.** (Example 7.12, page 467 in text.) For Example 7.6 find a 95% confidence interval for the mean improvement in the entire population of third-graders.

**Solution.** The **ToolPak** does not print a confidence interval directly, but the output provides enough information to carry out the calculations. Details are given in cells C18:E23 of Figure 7.12, which we have added to the output. The cells in column E of this block show the formulas that are the entries behind the cells in column D and whose values are evaluated and printed in the workbook by Excel. These formulas are the Excel equivalents of the formula

$$\bar{x}_1 - \bar{x}_2 \pm t^* \sqrt{\frac{s_1^2}{n_1} + \frac{s_2^2}{n_2}}$$

The information needed—the sample means, sample variances, sample sizes, and the critical $t^*$ values—is part of the **ToolPak** output and is referenced in cells C18:E23. A 95% confidence interval can then be read off as (1.23, 18.67).

## The Pooled Two-Sample $t$ Procedures

When the two populations are believed to be normal with the same variance, it is more common to use a pooled two-sample $t$ based on an exact $t$ distribution. The procedures are based on the statistic

$$t = \frac{(\bar{x}_1 - \bar{x}_2) - (\mu_1 - \mu_2)}{s_p\sqrt{\frac{1}{n_1} + \frac{1}{n_2}}} \tag{7.5}$$

where $s_p^2 = \frac{(n_1-1)s_1^2 + (n_2-1)s_2^2}{n_1+n_2-2}$ is called the pooled sample variance. This statistic is known to have an exact $t$ distribution on $\nu = n_1 + n_2 - 2$ degrees of freedom. The previous two-sample $t$ analyses carry over with the obvious modifications for the degrees of freedom and use of the denominator in (7.5) in place of the denominator in (7.2).

> **Example 7.8.** (Case 7.2 and Example 7.16, pages 476–478 in text.) In what ways are companies that fail different from those that continue to do business? To answer this question, one study compared various characteristics of 68 healthy and 33 failed firms. One of the variables considered was the ratio of current assets to current liabilities. Roughly speaking, this is the amount that the firm is worth divided by what it owes. The data appear in Table 7.4 on page 476 of the text and as the file *ta07004.dat* on the CD. At level $\alpha = 0.05$, test the hypothesis that there is no difference in this ratio between healthy and failed firms, that is test
>
> $$H_0 : \mu_1 = \mu_2$$
> $$H_a : \mu_1 \neq \mu_2$$

**Solution**

1. Enter the data and labels in A3:A71 (Healthy) and B3:B36 (Failed) of a workbook (Figure 7.13).

2. From the Menu Bar, choose **Tools – Data Analysis** and select **$t$-Test: Two-Sample Assuming Equal Variances** from the list of selections. (Refer to the dialog box in Figure 7.10.) Click **OK**.

3. Complete the next dialog box, which is similar to Figure 7.11, exactly as you did for the unequal variances case.

## Excel Output

The output appears in the range C3:E16 in Figure 7.13, and we see that

$$\bar{x}_1 = 1.7256 \qquad s_1^2 = 0.4087$$
$$\bar{x}_2 = 0.8236 \qquad s_2^2 = 0.2314$$
$$s_p^2 = 0.3514$$

The computed pooled $t$ statistic on 99 degrees of freedom is $t = 7.1721$ (cell D12), while the two-sided critical value at $\alpha = 0.05$ is $t^* = 1.9842$ (cell D16). The $P$-value is 0 to four significant digits. We conclude that there is very strong evidence that the ratio is higher for Healthy than for Failed firms.

| | A | B | C | D | E | F |
|---|---|---|---|---|---|---|
| 1 | | | Pooled Two-Sample t Test | | | |
| 2 | | | | | | |
| 3 | Healthy | Failed | t-Test: Two-Sample Assuming Equal Variances | | | |
| 4 | 1.50 | 0.82 | | | | |
| 5 | 0.10 | 0.89 | | Healthy | Failed | |
| 6 | 1.76 | 1.31 | Mean | 1.7256 | 0.8236 | |
| 7 | 1.14 | 0.05 | Variance | 0.4087 | 0.2314 | |
| 8 | 1.84 | 0.83 | Observations | 68 | 33 | |
| 9 | 2.21 | 0.90 | Pooled Variance | 0.3514 | | |
| 10 | 2.08 | 1.68 | Hypothesized Mean Difference | 0 | | |
| 11 | 1.43 | 0.99 | df | 99 | | |
| 12 | 0.68 | 0.62 | t Stat | 7.1721 | | |
| 13 | 3.15 | 0.91 | P(T<=t) one-tail | 0.0000 | | |
| 14 | 1.24 | 0.52 | t Critical one-tail | 1.6604 | | |
| 15 | 2.03 | 1.45 | P(T<=t) two-tail | 0.0000 | | |
| 16 | 2.23 | 1.16 | t Critical two-tail | 1.9842 | | |
| 17 | 2.50 | 1.32 | | | | |
| 18 | 2.02 | 1.17 | Mean Difference | 0.902 | =D6-E6 | |
| 19 | 1.44 | 0.42 | SE | 0.126 | =SQRT(D9)*SQRT((1/68+ 1/33)) | |
| 20 | 1.39 | 0.48 | t | 1.984 | =TINV(0.05, D11) | |
| 21 | 1.64 | 0.93 | ME | 0.250 | =D20*D19 | |
| 22 | 0.89 | 0.88 | lower | 0.652 | =D18-D21 | |
| 23 | 0.23 | 1.10 | upper | 1.151 | =D18+D21 | |
| 24 | 1.20 | 0.23 | | | | |
| 25 | 2.16 | 1.11 | | | | |
| 26 | 1.80 | 0.19 | | | | |
| 27 | 1.87 | 0.13 | | | | |
| 28 | 1.91 | 2.03 | | | | |
| 29 | 1.67 | 0.51 | | | | |
| 30 | 1.87 | 1.12 | | | | |
| 31 | 1.21 | 0.92 | | | | |
| 32 | 2.05 | 0.26 | | | | |
| 33 | 1.06 | 1.15 | | | | |
| 34 | 0.93 | 0.13 | | | | |
| 35 | 2.17 | 0.88 | | | | |
| 36 | 2.61 | 0.09 | | | | |
| 37 | 3.05 | | | | | |
| 38 | 1.52 | | | | | |
| 39 | 1.93 | | | | | |

Figure 7.13: Two-Sample $t$ ToolPak Output

## Confidence Intervals

**Example 7.8.** (Example 7.17, page 478 in text.) For Example 7.7 find a 95% confidence interval for the mean current assets to current liabilities ratio for Healthy vs. Failed firms.

**Solution.** Again, the **ToolPak** does not print a confidence interval directly, but it can be easily calculated using the summary statistics produced by the **ToolPak**. Details are given in cells C18:E23 of Figure 7.12 which we have added to the output. The cells in column E of this block show the formulas that are the entries behind the cells in column D¡ and whose values are evaluated and printed in the workbook by Excel. These formulas are the Excel equivalents of the formula

$$\bar{x}_1 - \bar{x}_2 \pm t^* s_p \sqrt{\frac{1}{n_1} + \frac{1}{n_2}}$$

The information needed—the sample means, sample variances, sample sizes, and the critical $t^*$ values—is part of the **ToolPak** output and is referenced in cells C18:E23. From D22:D23 we read a 90% confidence interval $(0.652, 1, 151)$.

## 7.3  Optional Topics in Comparing Distributions

## Inference for Population Spread

Suppose that $s_1^2$ and $s_2^2$ are the sample variances of independent simple random samples of sizes $n_1$ and $n_2$ taken from normal populations $N(\mu_1, \sigma_1)$ and $N(\mu_2, \sigma_2)$, respectively. Then the ratio

$$F = \frac{s_1^2/\sigma_1^2}{s_2^2/\sigma_2^2}$$

has a known sampling distribution that does not depend on $\{\mu_1, \mu_2, \sigma_1, \sigma_2\}$ but only on the sample sizes. It has an $F$ distribution on $n_1 - 1$ and $n_2 - 1$ degrees of freedom for the numerator and the denominator, respectively. The ratio on the right side of the equation is only one manifestation of the $F$ distribution, which is also used in analysis of variance and regression.

## The $F$ Test for Equality of Spread

In this section, the context is comparison of $\sigma_1$ and $\sigma_2$. It turns out for mathematical reasons that the appropriate parameter for testing the null hypothesis

$$H_0 : \sigma_1 \ = \ \sigma_2$$
$$H_a : \sigma_1 \ \neq \ \sigma_2$$

is the ratio $\frac{\sigma_1^2}{\sigma_2^2}$ rather than the difference, which we used for comparing means.

**Example 7.9.**  (Example 7.18, page 490 in text.) Determine whether it is appropriate to use the pooled $t$ test by testing equality of variances for the Healthy and Failed firms, using level $\alpha = 0.05$.

|   | A | B | C | D | E |
|---|---|---|---|---|---|
| 3 | Healthy | Failed | F-Test Two-Sample for Variances |  |  |
| 4 | 1.50 | 0.82 |  |  |  |
| 5 | 0.10 | 0.89 |  | Healthy | Failed |
| 6 | 1.76 | 1.31 | Mean | 1.726 | 0.824 |
| 7 | 1.14 | 0.05 | Variance | 0.409 | 0.231 |
| 8 | 1.84 | 0.83 | Observations | 68 | 33 |
| 9 | 2.21 | 0.90 | df | 67 | 32 |
| 10 | 2.08 | 1.68 | F | 1.77 |  |
| 11 | 1.43 | 0.99 | P(F<=f) one-tail | 0.0396 |  |
| 12 | 0.68 | 0.62 | F Critical one-tail | 1.89 |  |
| 13 | 3.15 | 0.91 |  |  |  |

Figure 7.14: $F$ Test Data and Output

**Solution.** We will use the $F$ test in the **Analysis ToolPak**.

1. Enter the data in columns A and B of a worksheet (Figure 7.14 shows part of this sheet.).

2. From the Menu Bar choose **Data Analysis – $F$-Test Two-Sample for Variances** and complete the dialog box of Figure 7.15. Notice that we have inserted not the specified level of significance $\alpha = 0.05$ in this box but rather half the value, 0.025, to reflect the fact that our test is two-sided, while the cells D11:D12 give the $P$-value and the critical value for a one-sided upper-tailed test.

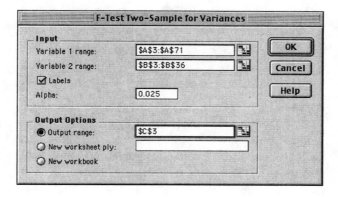

Figure 7.15: $F$ Test Dialog Box

**Caveat**

This tool requires that the larger of the two sample variances be in the numerator, so repeat this procedure by reversing the variables if the output shows that the variance of the data in **Variable 1 range** in Figure 7.15 is less than that in **Variable 2 range** (which is not the case here).

## Excel Output

The output appears in cells C6:E12, as shown in Figure 7.14. The computed value of $F$ under $H_0$

$$F = \frac{s_1^2}{s_2^2} = 1.77$$

appears in D10 (note that $s_1^2 > s_2^2$ as required) and does not exceed the 5% critical $F$ value 1.89 shown in D12.

We can obtain from cell D11 a one-sided $P$-value, which we need to double and find that the $P$-value $= 0.08$. Hence the standard deviations are significantly different at the 10% level but not the 5% level of significance.

## The $F$ Distribution Function

This a good place to record the syntax for the $F$ distribution. Suppose that $F$ is a random variable having an $F$ distribution with degrees of freedom $\nu_1$ for the numerator and $\nu_2$ for the denominator. Then for any $x > 0$,

$$\texttt{FDIST}(x, \nu_1, \nu_2) = P(F > x)$$

while for any $0 < p < 1$, the upper $p$ critical value is obtained from the inverse

$$P\left(F > \texttt{FINV}(p, \nu_1, \nu_2)\right) = p$$

# Chapter 8

# Inference for Proportions

Some statistical studies produce counts rather than measurements, for instance opinion polls that can be represented as 0 or 1-valued random variables. Sometimes there are more than two categories of counts. The parameters of interest for such data are the population proportions in the different categories.

In this chapter we discuss data representing the counts or proportions of outcomes occurring in different categories in a dichotomous population. The methods are very similar to large sample inferences on the mean of a population, with the main difference being in how the standard deviations are estimated.

## 8.1  Inference for a Single Proportion

To estimate the proportion $p$ of some characteristic in a population, it is common to take an SRS of size $n$ and count $X =$ the number in the sample possessing the characteristic. The sample proportion is

$$\hat{p} = \frac{X}{n}$$

For large $n$, the distribution of $X$ is approximately binomial $B(n, p)$, and by the central limit theorem the sample proportion

$$\hat{p} \text{ is approximately } N\left(p, \sqrt{\frac{p(1-p)}{n}}\right)$$

Inference is then based on the procedures for estimating a normal mean discussed in Chapter 6.

## Macro – Confidence Interval

Instead of using $\hat{p}$, we use a modification called the Wilson estimate, given by

$$\tilde{p} = \frac{X + 2}{n + 4}$$

The standard error of $\hat{p}$ is

$$\mathrm{SE}_{\hat{p}} = \sqrt{\frac{\tilde{p}(1 - \tilde{p})}{n + 4}}$$

where we have replaced $p$ with $\tilde{p}$ in the expression for the standard deviation of $\tilde{p}$. Therefore, a large-sample level $C$ confidence interval for $p$ is given as

$$\tilde{p} \pm z^* \mathrm{SE}_{\tilde{p}}$$

where $z^*$ is the upper $(1 - C)/2$ standard normal critical value.

> **Example 8.1.** (Example 8.1, page 507 in text.) Estimating the effect of work stress. The sample survey in Case 8.1 (page 504 in text) found that 68 of a sample of 100 employees agreed that work stress had a negative impact on their personal lives. Using the Wilson estimate, find a 95% confidence interval for the proportion of the restaurant chain's employees who feel that work stress is damaging their personal lives.

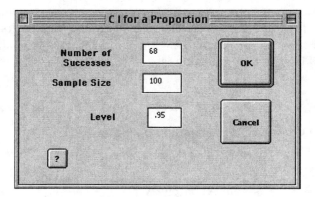

Figure 8.1: Macro Dialog Box – Confidence Interval for a Population Proportion

**Solution.**

1. Open the Excel file *1pCI.xls* and click the **Run** button to bring up the One-Sample proportion confidence interval dialog box and complete it as shown with the given data and the confidence level (Figure 8.1). Click **OK**.

2. The output appears on a new sheet in your workbook (Figure 8.2) from which you can read the answer (0.581, 0.785). These values differ in the third decimal place from the values in the text because of round off. The student may examine the formulas behind the entries in the output to learn how the Excel formulas are used.

| | A | B |
|---|---|---|
| 1 | **User Input** | |
| 2 | number of successes | 68 |
| 3 | sample size | 100 |
| 4 | confidence level | 0.95 |
| 5 | | |
| 6 | **Summary Statistics** | |
| 7 | sample proportion | 0.673 |
| 8 | | |
| 9 | **Calculations** | |
| 10 | standard error | 0.0469 |
| 11 | computed z | 1.9600 |
| 12 | margin of error | 0.0919 |
| 13 | | |
| 14 | **Confidence Limits** | |
| 15 | lower | 0.581 |
| 16 | upper | 0.765 |

Figure 8.2: Output –Confidence Interval for a Population Proportion

## Macro – Significance Test

For testing the null hypothesis

$$H_0 : p = p_0$$

we use the test statistic

$$z = \frac{\hat{p} - p_0}{\sqrt{\frac{p_0(1-p_0)}{n}}}$$

for a large-sample procedure. For example, if the alternative is two-sided

$$H_a : p \neq p_0$$

then we reject $H_0$ at level $\alpha$ if

$$|z| > z^*$$

where $z^*$ is the upper $\frac{\alpha}{2}$ standard normal critical value. The $P$-value is $2P(Z > |z|)$, where $Z$ is $N(0,1)$.

**Example 8.2.** (Example 8.2, page 510 in text.) A national survey of restaurant employees found that 75% said that work stress had a negative impact on their personal lives. A sample of 100 employees of a restaurant chain finds that 68 answer "Yes" when asked, "Does work stress have a negative impact on your personal life?" Is this good reason to think that the proportion of all employees of this chain who would say "Yes" differs from the national proportion $p_0 = 0.75$? To answer this question, test

$$H_0 : p = 0.75$$
$$H_a : p \neq 0.75$$

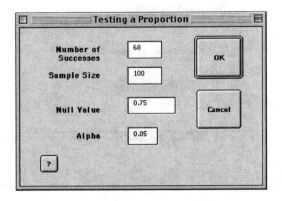

Figure 8.3: Macro Dialog Box – Significance Test for a Population Proportion

**Solution.**

1. Open the Excel file *1pT.xls* containing the significance test macro and click the **Run** button to bring up the macro dialog box. Complete it as shown in Figure 8.3 by entering the number of successes, the sample size, the null value, and the level of significance Alpha. Then click **OK**.

2. The output (Figure 8.4) shows three regions depending on the type of alternative. Ours is a two-sided test so the relevant portion gives a 5% critical value of 1.96, a computed $z$ value of $-1.62$, and a $P$-value of 0.106. The data are not significant and the conclusion is "Do not reject" $H_0$.

## 8.2  Comparing Two Proportions

Large-sample inference procedures for comparing the proportions $p_1$ and $p_2$ in two populations based on independent SRS of sizes $n_1$ and $n_2$, respectively, are also based on the normal approximation. These are common inference procedures. The natural estimate of the difference in proportions $D = p_1 - p_2$ is $\hat{D} = \hat{p}_1 - \hat{p}_2$. However, we will use the Wilson estimate, which differs very little for large samples. The Wilson estimates of $p_1$ and $p_2$ are

$$\tilde{p}_1 = \frac{X_1 + 1}{n_1 + 2} \quad \text{and} \quad \tilde{p}_2 = \frac{X_2 + 1}{n_2 + 2}$$

| | A | B |
|---|---|---|
| 1 | **User Input** | |
| 2 | null value | 0.75 |
| 3 | number of successes | 66 |
| 4 | sample size | 100 |
| 5 | alpha | 0.05 |
| 6 | | |
| 7 | **Summary Statistics** | |
| 8 | sample proportion | 0.66 |
| 9 | | |
| 10 | **Calculations** | |
| 11 | standard error | 0.04 |
| 12 | computed z | -1.62 |
| 13 | | |
| 14 | **Lower Test** | |
| 15 | critical value | -1.645 |
| 16 | decision | Do Not Reject H0 |
| 17 | P-value | 0.053 |
| 18 | | |
| 19 | **Upper Test** | |
| 20 | critical value | 1.645 |
| 21 | decision | Do Not Reject H0 |
| 22 | P-value | 0.947 |
| 23 | | |
| 24 | **Two-sided Test** | |
| 25 | critical value | 1.960 |
| 26 | decision | Do Not reject H0 |
| 27 | P-value | 0.106 |

Figure 8.4: Output – Significance Test for a Population Proportion

respectively, where $X_1$ and $X_2$ are the corresponding number of successes in the two samples. The corresponding Wilson estimate of $D$ is

$$\tilde{D} = \tilde{p}_1 - \tilde{p}_2$$

$\tilde{D}$ is approximately normal with mean $p_1 - p_2$ and standard deviation

$$\sigma_{\tilde{D}} = \sqrt{\frac{p_1(1-p_1)}{n_1} + \frac{p_2(1-p_2)}{n_2}}$$

## Confidence Intervals

We must replace the unknown parameters $p_1$ and $p_2$ by their estimates $\hat{p}_1$ and $\hat{p}_2$ to obtain an estimated standard error

$$\text{SE}_D = \sqrt{\frac{\hat{p}_1(1-\hat{p}_1)}{n_1} + \frac{\hat{p}_2(1-\hat{p}_2)}{n_2}}$$

and an approximate level $C$ confidence interval

$$\hat{p}_1 - \hat{p}_2 \pm z^* \text{SE}_D$$

where $z^*$ is the upper $(1-C)/2$ standard normal critical value.

**Example 8.3.**    (Example 8.5, page 524 in text.) The study in Case 8.3 suggested that there is a gender difference in the proportion of label users. Construct a confidence interval for the difference in proportions between women and men.

**Solution.**    Cells A4:C6 in Figure 8.5 summarize the data. Rather than use a macro, it is instructive to see the explicit calculations. These are shown on the worksheet. The values resulting from the calculations are given in column B, while the adjacent column C contains the formulas behind the values. Based on this data set, we are 95% confident that the difference in proportions between women and men is in the range (0.043, 0.165).

|   | A | B | C | D | E |
|---|---|---|---|---|---|
| 1 | | Confidence Interval for a Difference in Proportions | | | |
| 2 | | | | | |
| 3 | | Summary Statistics and User Input | | | |
| 4 | Group | n | X | X/n | (X+1)/(n+2) |
| 5 | women | 296 | 63 | 0.213 | 0.215 |
| 6 | men | 251 | 27 | 0.108 | 0.111 |
| 7 | | | | | |
| 8 | conf | 0.95 | | | |
| 9 | SE | 0.03101 | =SQRT((D5*(1-D5)/B5)+(D6*(1-D6)/B6)) | | |
| 10 | z | 1.960 | =NORMSINV(0.5+conf/2) | | |
| 11 | ME | 0.061 | =z*SE | | |
| 12 | Wilson estimate | 0.104 | =E5-E6 | | |
| 13 | Confidence Limits | | | | |
| 14 | lower | 0.043 | =(D5-D6)-ME | | |
| 15 | upper | 0.165 | =(D5-D6)+ME | | |

Figure 8.5: Confidence Interval—Difference in Proportions

## Significance Tests

The null hypothesis

$$H_0 : p_1 = p_2$$

is tested using the statistic

$$z = \frac{\hat{p}_1 - \hat{p}_2}{\text{SE}_{Dp}}$$

where $\text{SE}_{Dp}$ is the estimated standard deviation based on the pooled estimate

$$\hat{p} = \frac{X_1 + X_2}{n_1 + n_2}$$

of the common value $p \equiv p_1 = p_2$ of the population proportions. Here, $x_1$ and $x_2$ are the number of counts possessing the characteristic being counted in sample 1 and sample 2, respectively, and

$$\text{SE}_{Dp} = \sqrt{\hat{p}(1 - \hat{p}) \left( \frac{1}{n_1} + \frac{1}{n_2} \right)}$$

The decision rules based on $z$ are then analogous to those in Section 8.1. For example, if the alternative hypothesis is

$$H_a : p_1 > p_2$$

then we reject at level $\alpha$ if

$$z > z^*$$

where $z^*$ is the upper $\alpha$ standard normal critical value. Furthermore, the $P$-value is $P(Z > z)$, where $Z$ is $N(0,1)$.

**Example 8.4.** (Example 8.6, page 598 in text.) Referring to Example 8.3 above, test the hypothesis that women and men are equally likely to be label users.

**Solution.** The appropriate significance test is

$$H_0 : p_1 = p_2$$
$$H_a : p_1 \neq p_2$$

Figure 8.6 is a template that provides all the formulas required (located in the

| | A | B | C | D | E |
|---|---|---|---|---|---|
| **1** | | **Significance Test for the Difference in Proportions** | | | |
| **2** | | | | | |
| **3** | | Summary Statistics and User Input | | | |
| **4** | Group | n | X | X/n | |
| **5** | women | 296 | 63 | 0.213 | |
| **6** | men | 251 | 27 | 0.108 | |
| **7** | | | | | |
| **8** | null | 0 | Calculations | | |
| **9** | alpha | 0.05 | pooled_p | 0.165 | =(C5+C6)/(B5+B6) |
| **10** | alternate | upper | SE | 0.032 | =SQRT(pooled_p*(1-pooled_p)*(1/B5 + 1/B6)) |
| **11** | | | z | 3.309 | =((D5-D6)-null)/SE |
| **12** | Lower Test | | | | |
| **13** | lower_z | | =NORMSINV(alpha) | | |
| **14** | Decision | | =IF(z<lower_z,"Reject H0","Do Not Reject H0") | | |
| **15** | Pvalue | | =NORMSDIST(z) | | |
| **16** | Upper Test | | | | |
| **17** | upper_z | | =-NORMSINV(alpha) | | |
| **18** | Decision | | =IF(z>upper_z,"Reject H0","Do Not Reject H0") | | |
| **19** | Pvalue | | =1-NORMSDIST(z) | | |
| **20** | Two-Sided Test | | | | |
| **21** | two_z | 1.960 | =ABS(NORMSINV(alpha/2)) | | |
| **22** | Decision | Reject H0 | =IF(ABS(z)>two_z,"Reject H0","Do Not Reject H0") | | |
| **23** | Pvalue | 0.0009 | =2*(1-NORMSDIST(ABS(z))) | | |

Figure 8.6: Significance Test—Difference in Proportions

relevant cells in Columns C and E). Although this is a two-sided test, we have provided the formulas for lower and upper tests.

**Note:** We remind you that the formulas in Figure 8.5 and Figure 8.6 require **Named Ranges** to refer to the variables by their names, or else the cell references must be used.

Cell D9 gives the pooled sample proportion

$$\hat{p} = \frac{63 + 27}{296 + 251} = 0.165$$

The $z$ test statistic in cell D11 is 3.309, which is significant (cell B22), and the $P$-value in cell B23 is 0.0009.

# Chapter 9

# Inference for Two-Way Tables

In Example 8.4 we tested the equality of the responses of women and men to "No Sweat" labels on a garment. We were interested in comparing two populations (women, men) with respect to one response variable, "response to label." The response variable had two values, "yes" or "no." The data was displayed by the following table:

Table 9.1: No Sweat Labels – Comparing Two Proportions

| Population | $n$ | $X$ |
|---|---|---|
| 1 (women) | 296 | 63 |
| 2 (men) | 251 | 27 |
| Total | 547 | 90 |

To compare women and men, the sample proportions $\hat{p}_1$ and $\hat{p}_1$ were calculated from this table of counts and a test was carried out of the null hypothesis

$$H_0 : p_1 = p_2 \qquad (9.1)$$

where $p_1$ and $p_2$ are the proportions in the respective two populations who respond "yes."

There is another way to view and display the data. We may consider measuring two variables, gender and response, on the 547 individuals who were surveyed. When there are two possible levels for each variable, there are four combinations of measurements, and we could display the results as a frequency table or a bar graph involving four categories. But this would cause us to lose track of how the four categories are related to the levels and the variables (essentially the "geometry" of the design). It is more natural to display the data in a different (though equivalent) way than was presented in Table 9.1 by using a "two-dimensional" frequency table, shown here in Table 9.2.

Table 9.2: No Sweat Labels —2 × 2 Table

|            | Gender |       |       |
| Label user | Women  | Men   | Total |
|------------|--------|-------|-------|
| Yes        | 63     | 27    | 90    |
| No         | 233    | 224   | 457   |
| Total      | 296    | 251   | 547   |

This presentation of the data shows that there are two variables of interest, one that might be considered as an explanatory variable (gender) and the other as the response variable. The significance test of the null hypothesis in (9.1) judges whether there is a relationship between the two variables. If $p_1 = p_2$, then there is no relationship.

We wish to generalize this test to the situation in which there are more than two populations of interest or where the response variable can take more than two values.

A table such as Table 9.2 that shows data collected on two categorical variables having $r$ rows and $c$ columns of values for each of the two variables, is called an $r \times c$ table. In this chapter we discuss a technique based on the chi-square distribution for deciding if there is a relationship between two categorical variables.

## 9.1   Analysis of Two-Way Tables

We begin with an illustration using an Excel macro to analyse a 2 × 2 two-way table and then we will develop the methodology in general.

**Example 9.1.**   (Case 9.1, page 549 and Examples 9.3–9.5, pages 551–558 in text.) Many popular businesses are franchises. The relationship between the local entrepreneur and the franchise firm is spelled out in a detailed contract. One clause that the contract may or may not contain is the entrepreneur's right to an exclusive territory. How does the presence of an exclusive-territory clause in the contract relate to the survival of the business? A study designed to address this question collected data from a sample of 170 new franchise firms. Two categorical variables were measured for each firm – whether the firm was successful or not, and whether the contract offered an exclusive-territory clause. Here are the data:

| Successful | Exclusive territory | | Total |
|---|---|---|---|
| | Yes | No | |
| Yes | 108 | 15 | 123 |
| No | 34 | 13 | 47 |
| Total | 142 | 28 | 170 |

Use Excel to determine whether there is an association between classification by "success" and "classification by exclusive franchise." Specifically suppose that

$$\mathbf{p}_S = (p_{S1}, p_{S2})$$

represent the population proportions among the successful firms who had exclusive ($p_{11}$) or non-exclusive ($p_{12}$) contracts, and let

$$\mathbf{p}_N = (p_{N1}, p_{N2})$$

represent the corresponding proportions among the non-successful firms. The null hypothesis that is appropriate would be

$$H_0 : \mathbf{p}_S = \mathbf{p}_N \tag{9.2}$$

meaning equality of the vector of proportions.

## Macro – 2 × 2 Two-Way Table

**Solution.** Figure 9.1 shows the Excel file *chiSqR2x2.xls* containing a macro to test the null hypothesis 9.2 against a two-sided alternative. Enter the data exactly

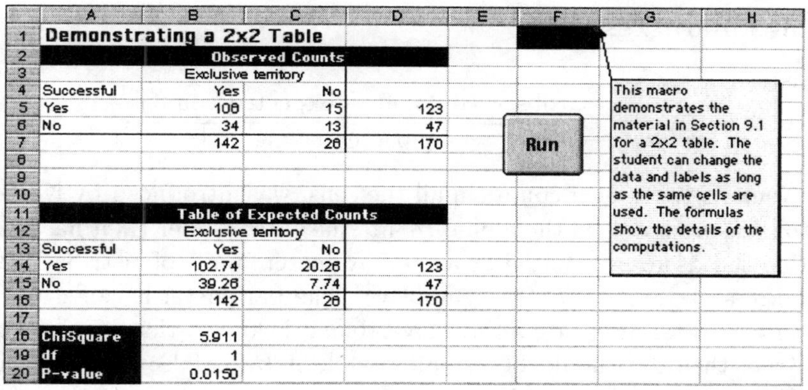

Figure 9.1: 2 × 2 Macro

as indicated into cells B5:C6 and add the row and column labels. You don't need to enter the row or column totals. The macro will put them in. When you click

**Run** a table of "Expected Counts" appears together with a chi-square statistic, the degrees of freedom, and the $P$-value. (Compare with the output in the text on pages 551–552). The small $P$-value $= 0.0150$ indicates that the proportions are not the same and that therefore there is an association between the two classifications.

## Expected Cell Counts

The table of expected values gives values which would be "expected" under the null hypothesis. The following provides the explanation for how these values are calculated. Under the null hypothesis, we may pool the successful and unsuccessful franchises into a total of $123 + 47 = 170$ franchises of which $108 + 34 = 142$ had exclusive territory. Therefore, under $H_0$, we can estimate the common value of $p_{S1} = p_{N1}$ by the pooled estimate

$$\frac{108 + 34}{123 + 47} = \frac{142}{170} = 0.835$$

to estimate the proportion having exclusive territory The row total for "Yes" under successful is 123. Thus, the number of observations in cell "successful franchise $\times$ exclusive territory" is a binomial random variable on 123 trials and having success probability estimated as 0.835. The expected number of counts $(n \times p)$ is thus

$$123 \times \frac{142}{170} = \frac{123 \times 142}{170} = 102.74$$

as shown in cell B5 in Figure 9.1. This leads to the useful mnemonic

$$\text{expected count} = \frac{\text{row total} \times \text{column total}}{\text{table total}}$$

Expected counts need to be computed for all cells.

## The Chi-Square Test

The statistic

$$X^2 = \sum \frac{(\text{observed count} - \text{expected count})^2}{\text{expected count}}$$

where the sum is taken over counts in all the cells, was introduced by Karl Pearson in 1900 to measure how well the model $H_0$ fits the data. Under $H_0$ it has a sampling distribution that is approximated by a one-parameter family of distributions known as chi-square and denoted by the symbol $\chi^2$. The parameter is called the degrees of freedom $\nu$, and for an $r \times c$ contingency table, it is known that $\nu = (r-1)(c-1)$. If $H_0$ is true, then $X^2$ should be "small," while if $H_0$ is false then $X^2$ should be "large." This leads to the criterion

Chi-square test:    Reject $H_0$ if $X^2 > \chi^2$

where $\chi^2$ is the upper critical $\alpha$ value of a chi-square distribution on $(r-1)(c-1)$ degrees of freedom. The value $X^2 = 5.911$ is shown in cell B16 and the degrees of freedom in cell B19 of Figure 9.1.

## The *P*-Value

The Excel function CHITEST calculates the *P*-value associated with the Pearson $X^2$ statistic. The syntax is

$$= \text{CHITEST}(\text{actual\_range}, \text{expected\_range})$$

where "actual_range" refers to the observed table of counts and "expected_range" refers to the expected table of counts. The *P*-value 0.0150 is computed by the macro in cell B18.

Refer to block F13:G14 of Figure 9.1 where we have given the formula and its value. The *P*-value is 0.000016, so we reject $H_0$ and conclude that there is an association between gender and sports goals.

## 9.2 Formulas and Models for Two-Way Tables

This section describes in detail all the calculations needed to carry out the above test. It generalizes the discussion in the previous section on expected cell counts and the chi-square test.

> **Example 9.2.** (Exercise 9.37, page 589 in text.) Knowing why different groups of customers participate in an activity or purchase a product can be very useful information in designing a marketing strategy. One study looked at why students participate in recreational sports and compared the profiles of men and women participants. One goal of

Table 9.3: Observed Counts for Sports Goals

| Goal | Gender | | |
| | Female | Male | Total |
|---|---|---|---|
| HSC-HM | 14 | 31 | 45 |
| HSC-LM | 7 | 18 | 25 |
| LSC-HM | 21 | 5 | 26 |
| LSC-LM | 25 | 13 | 38 |
| Total | 67 | 67 | 134 |

> people who participate in sports is social comparison – the desire to win or to do better than other people. Another is mastery – the desire to improve one's skills or to try one's best. Data were collected from 67 male and 67 female undergraduates at a large university. Each student was classified into one of four categories based on his or her responses to a questionnaire about sports goals. The four categories were high social comparison-high mastery (HSC-HM), high social comparison-low

mastery (HSC-LM), low social comparison-high mastery (LSC-HM), and low social comparison-low mastery (LSC-LM). One purpose of the study was to compare the goals of male and female students. The data are displayed in a two-way table (Table 9.3). The entries in this table are the observed, or sample, counts. For example, there are 14 females in the high social comparison-high mastery group. Determine whether there is an association between gender and goal by explicitly carrying out the calculations.

| | A | B | C | D | E | F | G |
|---|---|---|---|---|---|---|---|
| 1 | Computing Chi Square – Values | | | | | | |
| 2 | | | | | | | |
| 3 | | Observed Counts | | | | Chi-Square Cell Values | |
| 4 | | Female | Male | Row Total | | Female | Male |
| 5 | HSC-HM | 14 | 31 | 45 | | 3.2111 | 3.2111 |
| 6 | HSC-LM | 7 | 18 | 25 | | 2.4200 | 2.4200 |
| 7 | LSC-HM | 21 | 5 | 26 | | 4.9231 | 4.9231 |
| 8 | LSC-LM | 25 | 13 | 38 | | 1.8947 | 1.8947 |
| 9 | Column Total | 67 | 67 | 134 | | | |
| 10 | | | | | | Chi-Square | 24.898 |
| 11 | | Expected Counts | | | | Critical 5% value | 7.815 |
| 12 | | Female | Male | Row Total | | Decision | |
| 13 | HSC-HM | 22.5 | 22.5 | 45 | | Reject H0 | |
| 14 | HSC-LM | 12.5 | 12.5 | 25 | | | |
| 15 | LSC-HM | 13 | 13 | 26 | | | |
| 16 | LSC-LM | 19 | 19 | 38 | | | |
| 17 | Column Total | 67 | 67 | 134 | | | |

Figure 9.2: Computing the Value of $X^2$—Values

| | A | B | C | D | E | F | G |
|---|---|---|---|---|---|---|---|
| 1 | Computing Chi Square – Formulas | | | | | | |
| 2 | | | | | | | |
| 3 | | Observed Counts | | | | Chi-Square Cell Formulas | |
| 4 | | Female | Male | Row Total | | Female | Male |
| 5 | HSC-HM | 14 | 31 | 45 | | =(B5-B13)^2/B13 | =(C5-C13)^2/C13 |
| 6 | HSC-LM | 7 | 18 | 25 | | =(B6-B14)^2/B14 | =(C6-C14)^2/C14 |
| 7 | LSC-HM | 21 | 5 | 26 | | =(B7-B15)^2/B15 | =(C7-C15)^2/C15 |
| 8 | LSC-LM | 25 | 13 | 38 | | =(B8-B16)^2/B16 | =(C8-C16)^2/C16 |
| 9 | Column Total | 67 | 67 | 134 | | | |
| 10 | | | | | | Chi-Square | =SUM(F5:G8) |
| 11 | | Expected Counts | | | | Critical 5% value | =CHIINV(0.05, 3) |
| 12 | | Female | Male | Row Total | | Decision | |
| 13 | HSC-HM | 22.5 | 22.5 | 45 | | =IF(G10>G11, "Reject H0", "Do Not Reject H0") | |
| 14 | HSC-LM | 12.5 | 12.5 | 25 | | | |
| 15 | LSC-HM | 13 | 13 | 26 | | | |
| 16 | LSC-LM | 19 | 19 | 38 | | | |
| 17 | Column Total | 67 | 67 | 134 | | | |

Figure 9.3: Computing the Value of $X^2$—Formulas

**Solution.**

1. Enter the data into a worksheet as in block B4:C8 in Figure 9.2.

2. Copy the block B4:C8 to a convenient location, shown here copied to cells F4:G8. Change the label "Response" to "Chi-Square Cell Values."

3. The equation for $X^2$ is

$$X^2 = \sum \frac{(\text{observed count} - \text{expected count})^2}{\text{expected count}}$$

which we translate into Excel by the formula $= (\text{B5-B13})\hat{\ }2/\text{B13}$ entered in cell F5 and then filled to the block F5:G8. See Figure 9.3.

4. Sum all six cell entries by entering $= \text{SUM(F5:G8)}$ in cell G10. We get the value

$$X^2 = 24.898$$

The critical 5% $\chi^2$ value is given by $= \text{CHIINV(0.05,3)}$ in cell G11 and is

$$\chi^2_{.05} = 7.815$$

We therefore reject $H_0$, which is the same conclusion drawn using $P$-value.

You may arrange for the decision to appear on your workbook using the formula $= \text{IF}(G10 > G11, \text{"Reject H0"}, \text{"Do Not Reject H0"})$, which we have entered in cell G13.

**Note:** Excel has a function CHITEST that computes chi-square directly, although it does not provide the expected values.

# Chapter 10

# Inference for Regression

Predicting a response from explanatory variables is a fundamental objective in statistics. When the explanatory variables are quantitative, this means curve fitting. In this chapter we develop Excel methods for simple linear regression – fitting a straight line to one explanatory variable.

We pointed out in Chapter 2 that there are three procedures in Excel for regression analysis. These are complementary for the most part. We have already used the **Insert Trendline** command to fit and draw the least-squares regression line. In this chapter we derive additional information about the regression model with the **Regression** tool in the **ToolPak**. We also introduce some relevant Excel functions.

## 10.1   Inference about the Regression Model

**Example 10.1.** (Case 10.1, page 586 in text.) Many factors affect the wages of workers: the industry they work in, their type of job, their education and other experience, and changes in general levels of wages. Table 10.1, page 587 in the text, gives data for a sample of 59 married women who hold customer service jobs in Indiana banks, and shows their weekly wages at a specific point in time and also their length of service with their employer in months. Because industry, job type, and the time of measurement are the same for all 59 subjects, we expect to see a clear relationship between wages and length of service. Do a preliminary analysis by constructing a scatterplot and determining whether a straight line fit is appropriate.

**Solution**

1. Enter the data in two columns of a worksheet.

2. Using the **ChartWizard**, construct a scatterplot (Figure 10.1) of the data with LOS on the $x$-axis and Wages on the $y$-axis.

3. Select the scatterplot by clicking once within its region and from the Menu Bar choose **Chart − Add Trendline** .... Under the **Type** tab select linear and under the **Options** tab check the box to show the equation.

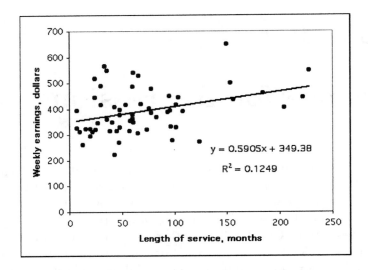

Figure 10.1: Scatterplot of Weekly Earnings vs. Length of Service

The scatterplot shows a moderate straight-line relationship with a regression equation

$$y = 349.38 + 0.590x$$

and a correlation coefficient $r = \sqrt{0.1249} = 0.3535$.

We did this already in Chapter 2. In this chapter we go beyond simple estimation of the regression line and develop inference procedures for the parameters that the slope and intercept represent, as well as for the underlying variation in the population from which the sample of 59 observation pairs was drawn.

## The Regression Tool

The general regression model for $n$ pairs of observations $(x_i, y_i)$ is

$$\text{DATA} = \text{FIT} + \text{RESIDUAL}$$

expressed mathematically as

$$y_i = \beta_0 + \beta_1 x_i + \varepsilon_i$$

The function

$$\mu_y = \beta_0 + \beta_1 x$$

is called the population (true) regression curve of $y$ on $x$, here taken to be linear. It represents the mean response of $y$ as a function of $x$. The quantities $\{\varepsilon_i\}$ are assumed to be independent normally distributed random variables, with a common mean 0 and common standard deviation $\sigma$. Therefore, there are three unknown parameters $(\beta_0, \beta_1, \sigma)$.

The parameters $\beta_0, \beta_1$ are estimated by the method of least-squares (described in Chapter 2) by the values $b_0, b_1$, respectively, where

$$b_0 = \bar{y} - b_1 \bar{x}$$
$$b_1 = r \frac{s_y}{s_x}$$

Recall that $s_x$ and $s_y$ are the sample standard deviations of the $x$ and $y$ data sets respectively.

The least-squares regression line, which estimates the population regression line, is given by the equation

$$\hat{y} = b_0 + b_1 x$$

Algebra shows that $b_0$ and $b_1$ are unbiased estimators of $\beta_0$ and $\beta_1$. We will look at their sampling distributions in Section 10.3.

In a manner entirely analogous to the case of estimating the variance of a data set, we obtain an estimate of the remaining parameter $\sigma^2$ as

$$s^2 = \frac{\sum_{i=1}^{n} e_i^2}{n-2}$$

where

$$e_i = y_i - \hat{y}_i$$

is the residual at $x_i$. The denominator $(n-2)$ makes the estimate unbiased.

Regression analysis thus consists not only of estimating $\beta_0, \beta_1$, and $\sigma^2$, but also making inferences about them and about predicted values corresponding to specified $x$-values. The primary tool is the **Regression Tool** in the Analysis ToolPak.

> **Example 10.2.** (Examples 10.2, page 592 in text.) Referring to the data in Example 10.1, use the **Regression Tool** to find the regression line and the estimated error $s$.

**Solution.** Although we have already obtained a scatterplot and the trendline from the **ChartWizard**, these results may also be obtained as options in the **Regression Tool**.

1. As before, enter the data in two columns of a worksheet with the independent variable $x$ to the left of the dependent variable $y$, for instance in columns A4:A63 and B4:B63, with the first row reserved for labels. (Excel requires this kind of data in contiguous regions.) Refer to Figure 10.3 showing the output of the **Regression Tool** with part of the data set.

2. From the Menu Bar choose **Tools – Data Analysis** and select **Regression** from the tools listed in the **Data Analysis** dialog box. Click **OK** to display the **Regression** dialog box (Figure 10.2).

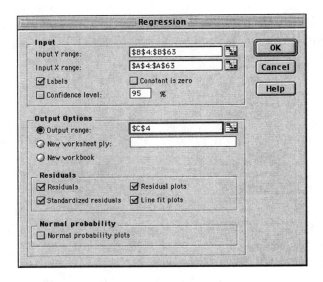

Figure 10.2: Regression Tool Dialog Box

3. Type the cell references B4:B63 in Figure 10.2 (or point and drag over the data in column B) for **Input Y range**. Do the same with respect to Column A for **Input X range**. Check the box **Labels**, leave **Constant is zero** clear because we are not forcing the line through the origin, and check the box **Confidence level** and insert "95." Under **Output options** select the radio button **Output range** and enter C4 to locate the upper left corner where the output will appear in the same workbook.

Check the following boxes:

**Residuals.**    To obtain predicted or fitted values $\hat{y}$ and their residuals.

**Residual Plots.**    For a scatterplot of the residuals against their $x$ values.

**Standardized Residuals.**    To obtain residuals divided by their standard error (useful to identify outliers).

**Line Fit Plots.**    To obtain a scatterplot of $y$ against $x$.

Do not check **Normal Probability Plots**.

**Note:** In making these selections, either use the Tab key to move from option to option or use the mouse. Click **OK**.

## Excel Output

The output is separated into six regions: Regression Statistics, ANOVA table, statistics about the regression line parameters, residual output, scatterplot with fitted line, and residual plot. We have reproduced a portion of the output in Figure 10.3. (This output also appears in the text at the top of page 592.) We next interpret the output in each of these regions.

| | A | B | C | D | E | F | G | H | I |
|---|---|---|---|---|---|---|---|---|---|
| 1 | **Regression Tool Example - Case 10.1** | | | | | | | | |
| 2 | | | | | | | | | |
| 3 | | | | | | | | | |
| 4 | LOS | Wages | SUMMARY OUTPUT | | | | | | |
| 5 | 94 | 389 | | | | | | | |
| 6 | 48 | 395 | *Regression Statistics* | | | | | | |
| 7 | 102 | 329 | Multiple R | 0.3535 | | | | | |
| 8 | 20 | 295 | R Square | 0.1249 | | | | | |
| 9 | 60 | 377 | Adjusted R Square | 0.1096 | | | | | |
| 10 | 78 | 479 | Standard Error | 82.2335 | | | | | |
| 11 | 45 | 315 | Observations | 59 | | | | | |
| 12 | 39 | 316 | | | | | | | |
| 13 | 20 | 324 | ANOVA | | | | | | |
| 14 | 65 | 307 | | *df* | *SS* | *MS* | *F* | *Significance F* | |
| 15 | 76 | 403 | Regression | 1 | 55034.359 | 55034.359 | 8.138 | 0.006 | |
| 16 | 48 | 378 | Residual | 57 | 385453.641 | 6762.345 | | | |
| 17 | 61 | 348 | Total | 58 | 440488 | | | | |
| 18 | 30 | 488 | | | | | | | |
| 19 | 108 | 391 | | *Coefficients* | *Standard Error* | *t Stat* | *P-value* | *Lower 95%* | *Upper 95%* |
| 20 | 61 | 541 | Intercept | 349.378 | 18.096 | 19.306 | 0.000 | 313.140 | 385.616 |
| 21 | 10 | 312 | LOS | 0.590 | 0.207 | 2.853 | 0.006 | 0.176 | 1.005 |

Figure 10.3: Data and Regression Tool Output

### Parameter Estimates

Rows 20 and 21 of the output provide statistics for the regression line parameters. From cell D20 we read the $y$-intercept $b_0 = 349.378$, and from cell D21 we read the slope of the regression line $b_1 = -0.590$. The regression line is therefore $\hat{y} = 349.38 + 0.590x$ as before. In cell D10 we read the standard error $s = 82.2335$. The remaining output we will ignore for the time being.

### Plots

Two plots are produced, a scatterplot of the data and a scatterplot of the residuals against $x$. Had we not used the **ChartWizard** in Example 10.1 to plot first the data, we could use the scatterplot produced by the regression tool. Figure 10.4 shows the scatterplot produced by Excel of the original data. By default, Excel uses markers even for the predicted value and does not show the line. Moreover, the scales need to be changed to produce a more useful graph. Although we already have a good scatterplot, it is valuable to show how to modify the Excel chart to a more useful form.

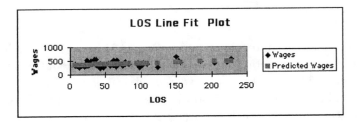

Figure 10.4: Default Scatterplot

## Changing Markers

First, we modify the Excel scatterplot by enlarging it. Then, we change the scale of the X and Y axes, replacing the diamond-shaped data plotting markers with circles. Finally, we replace the square-shaped predicted values with a line.

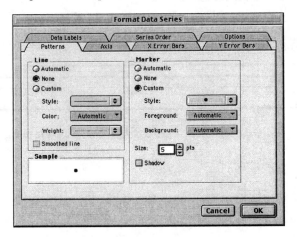

Figure 10.5: Formatting Markers

1. To resize the **Chart Area**, activate the Chart and drag one or more of the handles to the desired size. To resize the **Plot Area**, click within the plot area and drag one or more handles to the desired size.

2. When you resize the Chart, the scales on the X-axis and the Y-axis will also change. If the result is not satisfactory you can make these changes manually . To change the scale on the X-axis activate the Chart and double-click the X-axis. (Equivalently, click the X-axis once and choose **Format − Selected Axis...** from the Menu Bar.) Then under the **Scale** tab, change **Minimum** or **Maximum** to whatever is desired. The Y-axis may be edited similarly.

3. To change the data markers, first activate the Chart and select one of the data points. From the Menu Bar, choose **Format – Selected Data Series...**, click the **Patterns** tab, and select a **Custom Marker** (Figure 10.5).

## Changing Predicted Markers to a Line

The other important enhancement to the markers is to change the predicted ones in the default scatterplot to a line. This is a useful enhancement even in other contexts, and we separate its description here. The steps are as follows:

1. Activate the Chart and select one of the predicted markers. From the Menu Bar, choose **Format – Selected Data Series ...**.

2. In the **Format Data Series** dialog box, click the **Patterns** tab. Select radio button **Custom** for **Line**, and then pick a **Color**. Finally, select **None** for **Marker**. The markers disappear and are replaced by the regression line.

Finally, add titles and labels as described in Chapter 2. The resulting enhanced scatterplot should now resemble what was earlier produced by the **ChartWizard**.

## Significance Tests on the Slope and $y$ Intercept

All the information needed to carry out inference on the slope and intercept appear on the output. Suppose you wish to test

$$H_0 : \beta_1 = 0 \qquad \text{vs.} \qquad H_a : \beta_1 > 0$$

because we expect that wages will rise with the length of service. The appropriate test statistic

$$t = \frac{b_1}{SE_{b_1}}$$

where $SE_{b_1}$ is the standard error of $b_1$ and is obtained from the standard deviation of $b_1$ by replacing the unknown $\sigma$ with $s$. Its value is given in the output. In cell E21 in Figure 10.3 we read $SE_{b_1} = 0.207$. The $t$ statistic is also given in F21, denoted as $t$ stat. It is

$$t = \frac{b_1}{SE_{b_1}} = \frac{0.590}{0.0.207} = 2.853$$

The corresponding two sided $P$-value appears in cell G21. Because our test is one-sided, we take half that value,

$$P\text{-value} = \frac{1}{2}(0.006) = 0.003$$

As you can see, we don't need to know how $SE_{b_1}$ is obtained to carry out this test. The form of $SE_{b_1}$ will be discussed in Section 10.3.

Similarly, for testing

$$H_0 : \beta_0 = 0 \qquad \text{vs.} \qquad H_a : \beta_0 \neq 0$$

the appropriate test statistic is

$$t = \frac{b_0}{\text{SE}_{b_0}}$$

where $\text{SE}_{b_0}$ estimates the true standard deviation $\sigma_{b_0}$ of $b_0$. From D20:G20 we read the relevant estimates and the computed test statistic

$$t = \frac{b_0}{\text{SE}_{b_0}} = \frac{349.378}{18.0896} = 19.306$$

together with the two-sided

$$P\text{-value} = 0.0000$$

### Confidence Intervals for the Slope and $y$ Intercept

We used the default of 95% for the confidence level when we completed the **Regression** tool dialog box. The lower and upper 95% confidence limits appear in H20:I21 in Figure 10.3. Thus 95% confidence intervals arePBS

$$\text{for } \beta_0 \qquad (313.14, 385.616)$$
$$\text{for } \beta_1 \qquad (0.176, 1.005)$$

## 10.2   Inference about Prediction

In the preceding section we drew inferences on the parameters $\beta_0$ and $\beta_1$ in the regression line $\mu_y = \beta_0 + \beta_1 x$. Here, we examine the mean response and the prediction of a single outcome, both at a specified value $x^*$ of the explanatory variable.

- To estimate the mean response, we use a confidence interval for the parameter $\mu_y$ based on the point estimate

$$\hat{\mu}_y \equiv \hat{y} = b_0 + b_1 x$$

A level $C$ confidence interval for $\mu_y$ is given by

$$\hat{y} \pm t^* \text{SE}_{\hat{\mu}} \tag{10.1}$$

where the standard error is given by

$$\text{SE}_{\hat{\mu}} = s\sqrt{\frac{1}{n} + \frac{(x^* - \bar{x})^2}{\sum_{i=1}^{n}(x_i - \bar{x})^2}}$$

- To predict a single observation at $x^*$, a level $C$ prediction interval given by

$$\hat{y} \pm t^* \text{SE}_{\hat{y}} \tag{10.2}$$

where the appropriate standard error is now given by

$$\text{SE}_{\hat{y}} = s \sqrt{1 + \frac{1}{n} + \frac{(x^* - \bar{x})^2}{\sum_{i=1}^{n}(x_i - \bar{x})^2}}$$

In each case, $t^*$ is the upper $(1 - C)/2$ critical value of the Student $t$ distribution with $n - 2$ degrees of freedom.

Excel's **Regression Tool** does not provide either of these intervals but does provide the output needed (namely $b_0, b_1$, and $s$, from which $\hat{y}$ and $\text{SE}_{\hat{\mu}}$ can be calculated) to obtain both and (10.1) and (10.2).

### Prediction Interval for a Future Observation

**Example 10.3.** (Examples 10.5, page 610 in text.) Tonya has worked in customer service at an Indiana bank for 125 months. We don't know what she earns, but we can use the data on other bank workers to give a 95% prediction interval for her earnings.

### Solution.

1. First, the **Regression Tool** needs to be invoked to determine the regression line. We did this already in Example 10.1, so refer to the output, duplicated in Figure 10.6, which also shows the additional computations for obtaining the prediction intervals (and confidence intervals for Example 10.4 below). If you asked for residuals when using the **Regression Tool**, then move or delete them to make room in your sheet.

2. The value of $b_0$ is in cell D20, $b_1$ is in D21, and $s$ is in D10. The regression line is therefore $\hat{y} = 349.378 + 0.590x$.

3. Referring to Figure 10.6, type the formula

$$=\text{D20}+\text{D21}*\text{D25}$$

into cell D26, giving a predicted (fit) value $\hat{y} = 423.18$. In cell B27, enter the formula

$$= \text{D10} * \text{SQRT}(1 + 1/59 + (\text{D25} - \text{AVERAGE}(\text{A5:A63}))\hat{}2/(59 * \text{VARP}(\text{A5:A63})))$$

giving $\text{SE}_{\hat{y}} = 83.65$ (called StDev Fit in the worksheet). In D28, enter the formula

$$= \text{D29} - \text{TINV}(0.05, 57) * \text{D27}$$

| | A | B | C | D | E | F | G | H | I |
|---|---|---|---|---|---|---|---|---|---|
| 1 | **Regression Tool Example – Case 10.1** | | | | | | | | |
| 2 | | | | | | | | | |
| 3 | | | | | | | | | |
| 4 | LOS | Wages | SUMMARY OUTPUT | | | | | | |
| 5 | 94 | 389 | | | | | | | |
| 6 | 48 | 395 | *Regression Statistics* | | | | | | |
| 7 | 102 | 329 | Multiple R | 0.3535 | | | | | |
| 8 | 20 | 295 | R Square | 0.1249 | | | | | |
| 9 | 60 | 377 | Adjusted R Square | 0.1096 | | | | | |
| 10 | 78 | 479 | Standard Error | 82.2335 | | | | | |
| 11 | 45 | 315 | Observations | 59 | | | | | |
| 12 | 39 | 316 | | | | | | | |
| 13 | 20 | 324 | ANOVA | | | | | | |
| 14 | 65 | 307 | | *df* | *SS* | *MS* | *F* | *Significance F* | |
| 15 | 76 | 403 | Regression | 1 | 55034.359 | 55034.359 | 8.138 | 0.006 | |
| 16 | 48 | 378 | Residual | 57 | 385453.641 | 6762.345 | | | |
| 17 | 61 | 348 | Total | 58 | 440488 | | | | |
| 18 | 30 | 488 | | | | | | | |
| 19 | 108 | 391 | | *Coefficients* | *Standard Error* | *t Stat* | *P-value* | *Lower 95%* | *Upper 95%* |
| 20 | 61 | 541 | Intercept | 349.378 | 18.096 | 19.306 | 0.000 | 313.140 | 385.616 |
| 21 | 10 | 312 | LOS | 0.590 | 0.207 | 2.853 | 0.006 | 0.176 | 1.005 |
| 22 | 68 | 418 | | | | | | | |
| 23 | 54 | 417 | | | | | | | |
| 24 | 24 | 516 | Prediction Interval for a Future Observation | | | | | | |
| 25 | 222 | 443 | future x | 125 | | | | | |
| 26 | 58 | 353 | Fit | 423.18 | =D20+D21*D25 | | | | |
| 27 | 41 | 349 | StDev Fit | 83.65 | =D10*SQRT(1+1/59+(D25-AVERAGE(A5:A63))^2/(59*VARP(A5:A59) | | | | |
| 28 | 153 | 499 | PI lower limit | 255.67 | =D26-TINV(0.05,57)*D27 | | | | |
| 29 | 16 | 322 | PI upper limit | 590.70 | =D26+TINV(0.05,57)*D27 | | | | |
| 30 | 43 | 408 | | | | | | | |
| 31 | 96 | 393 | | | | | | | |
| 32 | 98 | 277 | Confidence Interval for Mean Response | | | | | | |
| 33 | 150 | 649 | future x | 125 | | | | | |
| 34 | 124 | 272 | Fit | 423.18 | =D20+D21*D25 | | | | |
| 35 | 60 | 486 | StDev Fit | 15.35 | =D10*SQRT(1/59+(D33-AVERAGE(A5:A63))^2/(59*VARP(A5:A59))) | | | | |

Figure 10.6: Confidence and Prediction Intervals

and in B29, enter the formula

$$= B23 + \text{TINV}(0.05, 57) * D27$$

These formulas are the Excel equivalents of (10.2). (For convenience, we have shown these formulas on the workbook. The formulas are in the block D26:D29 and the values are in D26:D29.) Read the 95% prediction interval (255.67, 590.70) in D28:D29 which coincides with the interval given in the text in the middle of page 610 and the middle of page 617.

### Confidence Interval for a Mean

Entirely analogous steps are used to derive a confidence interval for the mean value of the regression line at a specified value of $x$.

**Example 10.4.** (Exercise 10.25, page 610 in text.) Find a 95% confidence for the mean earnings of all bank workers with 125 months of service.

**Solution.** The only difference between this question and the previous example lies in the value of the appropriate standard error $SE_\mu$. All formulas are shown in Figure 10.6, and from cells D36:D37 we read (392.45, 453.92).

## Regression Functions

An alternative to the **Regression Tool** uses Excel regression functions. This is the third method mentioned in Chapter 2 for regression analysis.

### FORECAST

**Example 10.5.** Use the FORECAST function to predict Tonya's weekly earnings.

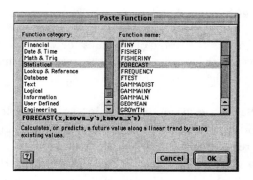

Figure 10.7: Paste Function

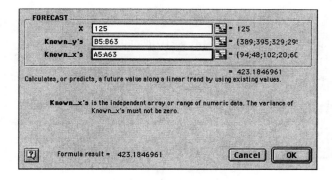

Figure 10.8: The FORECAST Dialog Box

**Solution** Without having to derive the regression line and calculating

$$\hat{y} = 349.38 + 0.590(125) = 423.18$$

we could use the FORECAST function, whose syntax is

$$\text{FORECAST}(x, \text{Known\_y's}, \text{Known\_x's})$$

Here **Known_x's** refers to the set $\{x_i\}$ while **Known_y's** refers to the $\{y_i\}$. The function can be called from the **Formula Palette** or the **Function Wizard**, in which case you should select **Statistical** for **Function category**, FORECAST for **Function name** (Figure 10.7), and insert the required parameters in the dialog box to obtain the predicted value shown at the bottom of Figure 10.8.

## TREND

A more general method for obtaining predicted values is TREND with syntax

$$\text{TREND}(\text{Known\_y's}, \text{Known\_x's}, \text{New\_x's}, \text{Const})$$

The parameter **New_x's** is the range of $x$-values for which predictions are desired. The parameter **Const** determines whether the regression line is forced through the origin ($\beta_0 = 0$). Use the value 1 for general $\beta_0$.

> **Example 10.6.** As in many other businesses, technological advances and new methods have produced dramatic results in agriculture. Cells A3:B7 in Figure 10.6 give the data on the average yield in bushels per acre of U.S. corn for selected years. Here, year is the explanatory variable $x$, and yield is the response variable $y$. A scatterplot suggests that we can use linear regression to model the relationship between yield and time. Predict corn yields in the years 2000, 2001, 2002, 2003, and 2004.

| D4 | | | = | {=TREND(B4:B7,A4:A7,C4:C8)} | |
|----|----|----|----|----|----|
| | A | B | C | D | E |
| 1 | | Using the TREND Function | | | |
| 2 | | | | | |
| 3 | Year | Yield | x | predicted | |
| 4 | 1966 | 73.1 | 2000 | 138.646 | |
| 5 | 1976 | 88.0 | 2001 | 140.58 | |
| 6 | 1986 | 119.4 | 2002 | 142.514 | |
| 7 | 1996 | 127.1 | 2003 | 144.448 | |
| 8 | | | 2004 | 146.382 | |

Figure 10.9: The TREND Function

**Solution.**

1. Select a region of cells D4:D8 (Figure 10.9) for the output.

2. Click the **Formula Palette**, select the `TREND` function, and enter the values B4:B7, A4:A7, and C4:C8 into the argument fields **Known_*y*'s**, **Known_*x*'s**, and **New_*x*'s**, respectively. Leave the **Const** field blank. Click **OK**.

3. Click the mouse pointer in the **Formula Bar**, hold down the **Shift** and **Control** keys (either **Macintosh** or **Windows**), and press **Enter** to **Array-Enter** the formula. The formula will appear surrounded by braces in the Formula Bar, indicating that it has been array-entered, and the predicted values {138.648, 140.580, 142.514, 144.448, 146.382} will appear in the output range (Figure 10.9).

## LINEST

The last regression function described here is `LINEST`, which returns the estimated

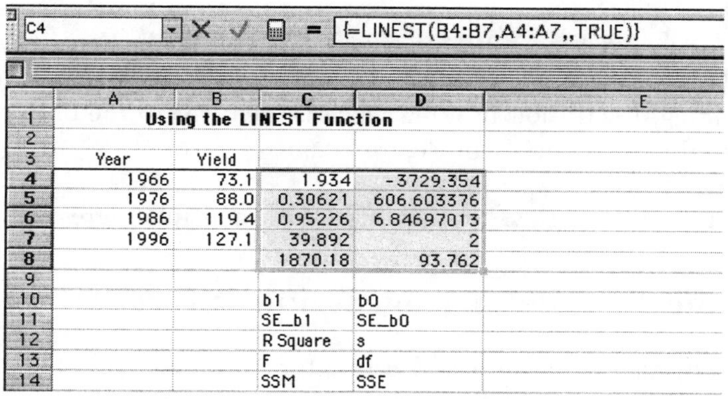

C4 = {=LINEST(B4:B7,A4:A7,,TRUE)}

| | A | B | C | D | E |
|---|---|---|---|---|---|
| 1 | Using the LINEST Function | | | | |
| 2 | | | | | |
| 3 | Year | Yield | | | |
| 4 | 1966 | 73.1 | 1.934 | -3729.354 | |
| 5 | 1976 | 88.0 | 0.30621 | 606.603376 | |
| 6 | 1986 | 119.4 | 0.95226 | 6.84697013 | |
| 7 | 1996 | 127.1 | 39.892 | 2 | |
| 8 | | | 1870.18 | 93.762 | |
| 9 | | | | | |
| 10 | | | b1 | b0 | |
| 11 | | | SE_b1 | SE_b0 | |
| 12 | | | R Square | s | |
| 13 | | | F | df | |
| 14 | | | SSM | SSE | |

Figure 10.10: The `LINEST` Function

least-squares line coefficients, their standard errors, $r^2$, $s$, the computed $F$ statistic with its degrees of freedom, and the regression and error sums of squares. The syntax is

$$\texttt{LINEST}(x, \text{Known_}y\text{'s}, \text{known_}x\text{'s}, \text{Const}, \text{Stats})$$

If the field **Stats** is set to false, only the regression coefficients are output; true will produce an entire block of output. The default is false.

> **Example 10.7.** (Example 10.6 continued.) Use `LINEST` to perform a regression analysis of corn yield versus time for the data in Figure 10.6.

### Solution

1. Select a block of cells C4:D8 with two columns and five rows. (For multiple regression you require a block in which the number of columns equals the number of independent variables plus one.)

2. Click the **Formula Palette**, select the LINEST function, and enter the values B4:B7 and A4:A7 into the argument fields **Known_y's** and **Known_x's**, respectively. Leave the **Const** field blank and type true in the **Stats** field. Click **OK**.

3. The formula = LINEST(B4:B7,A4:A7) will be visible in the **Formula Bar**, but only the value 1.93400 in cell C4 will appear in your sheet. With the block C4:D8 still selected, click the mouse pointer in the **Formula Bar**, as you did for the TREND function, hold down the **Shift** and **Control** keys, and press **Enter** to **Array-Enter** the formula. The formula will appear surrounded by braces in the **Formula Bar**, indicating that it has been array-entered, and the output will appear in C4:C8 (Figure 10.10).

In cells C10:D14 in Figure 10.10, we have described the contents of the values in C4:D8. These may be compared with the output from the **Regression** tool for this data set, shown in Figure 10.6. For instance, cell D8 in Figure 10.10 contains the value 93.76200. This represents the error sum of squares SSE, as seen from the corresponding description in D14, which is identical to the output in cell E15 of the **Regression** tool output in Figure 10.6.

**Note:** Both TREND and LINEST can be used with multiple regression.

## 10.3    Some Details of Regression Inference

The statistics $b_0$ and $b_1$ are unbiased estimators of $\beta_0$ and $\beta_1$. They are random variables having sampling distributions:

(i) $b_0$ is normal with

$$\text{mean} \quad = \quad \mu_{b_0} = \beta_0$$

$$\text{standard deviation} \quad = \quad \sigma_{b_0} = \sigma\sqrt{\frac{1}{n} + \frac{\bar{x}^2}{\sum_{i=1}^{n}(x_i - \bar{x})^2}}$$

(ii) $b_1$ is normal with

$$\text{mean} \quad = \quad \mu_{b_1} = \beta_1$$

$$\text{standard deviation} \quad = \quad \sigma_{b_1} = \frac{\sigma}{\sqrt{\sum_{i=1}^{n}(x_i - \bar{x})^2}}$$

(iii) The estimate $s^2 = \frac{\sum_{i=1}^{n}(y_i - \hat{y}_i)_i^2}{n-2}$ of the population variance $\sigma^2$ has the following sampling property:

$$\frac{(n-2)s^2}{\sigma^2} \text{ is chi-squared on } (n-2) \; df$$

This means that the standard error of the slope $b_1$ of the regression line is

$$\text{SE}_{b_1} = \frac{s}{\sqrt{\sum_{i=1}^{n}(x_i - \bar{x})^2}}$$

while the standard error of the $y$-intercept $b_0$ is

$$\text{SE}_{b_0} = s\sqrt{\frac{1}{n} + \frac{(x^* - \bar{x})^2}{\sum_{i=1}^{n}(x_i - \bar{x})^2}}$$

## Analysis of Variance for Regression

In this section, we consider the remaining two elements of the output from the **Regression Tool**. This material is discussed on pages 618–622 in the text.

### ANOVA Table

The ANOVA approach uses an $F$ test to determine whether a substantially better fit is obtained by the regression model than by a model with $\beta_1 = 0$. The ANOVA output breaks the observed total variation in the data

$$\text{SST} = \sum_{i=1}^{n}(y_i - \bar{y})^2$$

into two components, residual or error sum of squares

$$\text{SSE} = \sum_{i=1}^{n}(y_i - \hat{y}_i)^2$$

and a model sum of squares

$$\text{SSM} = \sum_{i=1}^{n}(\hat{y}_i - \bar{y})^2$$

connected by the identity

$$\text{SST} = \text{SSE} + \text{SSM}$$

The test criterion is based on how much smaller the residual sum of squares is under each fit and is based on the $F$ ratio

$$F = \frac{\text{MSM}}{\text{MSE}}$$

where $\text{MSM} = \frac{\text{SSM}}{1}$ is the mean square for the model fit, while $\text{MSE} = \frac{\text{SSE}}{n-2}$ is the mean square for error. Under the null hypothesis that $\beta_1 = 0$, this ratio has an $F$ distribution, with 1 degree of freedom in the numerator and $n - 2$ degrees of freedom in the denominator.

All the above calculations appear in rows 13–16 in Figure 10.11. This is an advanced topic, and the reader is referred to the text for a more complete discussion. The only point we add here is that the $F$ test is identical to a two-sided test for $H_0 : \beta_1 = 0$ vs. $H_a : \beta_1 \neq 0$, and the $F$ statistic 8.138 in cell G15 is always the square of the $t$ stat = 2.853 appearing in cell F21 in Figure 10.3 in this context.

| | C | D | E | F | G | H |
|---|---|---|---|---|---|---|
| 13 | ANOVA | | | | | |
| 14 | | df | SS | MS | F | Significance F |
| 15 | Regression | 1 | 55034.359 | 55034.359 | 8.138 | 0.006 |
| 16 | Residual | 57 | 385453.641 | 6762.345 | | |
| 17 | Total | 58 | 440488 | | | |

Figure 10.11: ANOVA Output

## Regression Statistics

| | C | D |
|---|---|---|
| 6 | *Regression Statistics* | |
| 7 | Multiple R | 0.3535 |
| 8 | R Square | 0.1249 |
| 9 | Adjusted R Square | 0.1096 |
| 10 | Standard Error | 82.2335 |
| 11 | Observations | 59 |

Figure 10.12: Regression Statistics Output

This is the last component of the output from Figure 10.3, which we discuss briefly and isolate in Figure 10.12, which gives, for instance, the coefficient of determination $r^2 = 0.1249$ and the standard error $s = 82.2335$, in addition to more advanced statistics such as the Adjusted $R$ Square used in multiple regression. These formulas may be used directly to carry out inference given by Excel.

**Example 10.8.** (Example 10.7, page 619 in text.) Verify directly from the sums of squares given in the ANOVA output in Figure 10.11 that the coefficient of determination $r^2 = 0.1249$.

**Solution.** From cells E15 and E17

$$
\begin{aligned}
r^2 &= \frac{\text{Regression Sum of Squares}}{\text{Total Sum of Squares}} \\
&= \frac{55034.36}{440488} \\
&= 0.1249
\end{aligned}
$$

which is shown in cell D8 as part of the "Regression Statistics" output.

**Example 10.9.** (Example 10.8, page 621 in text.) Verify directly from the sums of squares given in the ANOVA output in Figure 10.11 that the $F$ statistic is 8.138. Then check that the square root of $F$ is equal to the value of the $t$ statistic, which appears in cell F10 in Figure 10.3.

**Solution.**    From cells F15 and F16

$$F = \frac{\text{Regression Mean Square}}{\text{Residual Mean Square}}$$
$$= \frac{55034.36}{6732.345}$$
$$= 8.138$$

which is shown in cell G15. Also $\sqrt{8.138} = 2.853$, which is the $t$ statistic used to test whether the slope of the regression line is 0.

# Chapter 11

# Multiple Regression

Multiple regression extends simple linear regression by fitting surfaces to data involving two or more explanatory variables $\{x_1, \ldots, x_p\}$ used to predict a response variable $y$. For example,

> Table 11.1, page 635 in the text, shows some characteristics of thirty stocks in the Dow Jones Industrial Average. Included are the stock symbol, company name, assets (in billions of dollars) for the year 1999. How are profits related to sales and assets? In this case, profits represent the response variable, and sales and assets are two explanatory variables.

## 11.1 Data Analysis for Multiple Regression

The data consist of $n$ sets of observations

$$(x_{i1}, x_{i2}, \ldots, x_{ip}, y_i) \qquad 1 \leq i \leq n$$

where $y_i$ is the response and $(x_{i1}, \ldots, x_{ip})$ are the values of the $p$ explanatory variables measured on the $i$th subject. Generalizing Chapter 10, we wish to fit a linear model

$$\hat{y}_i = b_0 + b_1 x_{i1} + b_2 x_{i2} + \cdots + b_p x_{ip}$$

predicting the response from the explanatory variables. The coefficients $\{b_1, \ldots, b_p\}$ are estimated by the Principle of Least-Squares described in Section 10.1. The regression coefficient $\beta_i$ measures how much the response changes if $x_i$ changes one unit, keeping all other explanatory variables fixed. The $i$th residual is

$$e_i = y_i - \hat{y}_i$$

For example, in the situation described above, $n = 30$, the subjects are the firms, and $p = 2$.

$$x_1 = \text{assets}$$
$$x_1 = \text{sales}$$
$$y = \text{profits}$$

| | A | B | C | D |
|---|---|---|---|---|
| 1 | **Dow Jones 30 Industrials: Assets, Sales, Profits** | | | |
| 2 | | | | |
| 3 | Firm | Assets | Sales | Profits |
| 4 | Alcoa | 17.07 | 16.45 | 1.05 |
| 5 | AmericanExpress | 148.52 | 21.28 | 2.48 |
| 6 | AT&T | 169.41 | 62.39 | 3.43 |
| 7 | Boeing | 36.15 | 57.99 | 2.31 |
| 8 | Caterpillar | 26.64 | 19.70 | 0.95 |
| 9 | Citigroup | 716.94 | 82.01 | 9.87 |
| 10 | Coca-Cola | 21.62 | 19.81 | 2.43 |
| 11 | WaltDisney | 43.68 | 23.75 | 1.30 |
| 12 | DuPont | 40.78 | 16.91 | 7.69 |
| 13 | EastmanKodak | 14.37 | 14.09 | 1.39 |
| 14 | ExxonMobil | 144.52 | 130.97 | 7.91 |
| 15 | GeneralElectric | 405.20 | 110.83 | 10.72 |
| 16 | GeneralMotors | 274.73 | 167.37 | 6.00 |
| 17 | HomeDepot | 35.30 | 42.11 | 3.49 |
| 18 | Honeywell | 13.47 | 38.43 | 2.32 |
| 19 | Hewlett-Packard | 23.53 | 23.74 | 1.54 |
| 20 | InternationalBusines | 43.85 | 29.39 | 7.31 |
| 21 | Intel | 87.50 | 87.55 | 7.71 |
| 22 | InternationalPaper | 30.27 | 24.94 | 0.18 |
| 23 | J.P.MorganChase | 29.16 | 27.47 | 4.17 |
| 24 | Johnson&Johnson | 20.98 | 13.26 | 1.95 |
| 25 | McDonald's | 35.63 | 32.71 | 5.89 |
| 26 | Merck | 37.16 | 21.86 | 7.79 |
| 27 | Microsoft | 13.90 | 15.66 | 1.76 |
| 28 | MinnesotaMining&M | 260.90 | 18.07 | 2.06 |
| 29 | PhilipMorris | 61.38 | 61.75 | 7.68 |
| 30 | Procter&Gamble | 32.11 | 39.19 | 3.76 |
| 31 | SBCCommunications | 83.21 | 49.49 | 8.16 |
| 32 | UnitedTechnologies | 24.37 | 23.84 | 1.53 |
| 33 | Wal-Mart | 70.25 | 165.01 | 5.38 |

Figure 11.1: Dow Jones Data Set

Excel uses the **Regression Tool** for multiple regression. The data are entered into a workbook, one row for each observation, so that the columns correspond to the explanatory variables, followed by the response variable (all in adjacent columns). As in Chapter 10, the output is separated into six regions: regression statistics, ANOVA table, statistics about parameters, residuals, scatterplots, and residual plots. However, before using this tool, we will do some preliminary descriptive data analysis.

## Preliminary Exploratory Data Analysis

In this complex setting with several explanatory variables, it is important to use some of the descriptive tools described in Chapters 1 and 2 as preliminary steps to obtain numerical and graphical summaries first.

**Example 11.1.** (Example 11.2, page 636 and continued through pages 647 of text.) Use Excel to obtain individual descriptive statistics, correlations, and histograms for the observations.

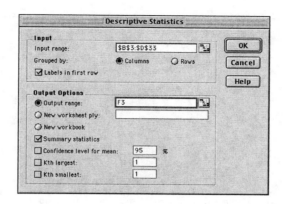

Figure 11.2: Descriptive Statistics Dialog Box

**Solution.**

1. Import the data using the **Text Import Wizard** into columns of an Excel worksheet as in Figure 11.1.

2. Next, go to the Menu Bar and choose **Tools – Data Analysis – Descriptive Statistics** and complete it as in Figure 11.2. The output which appears in your worksheet is shown in Figure 11.3.

3. Now choose **Tools – Data Analysis – Histogram** from the Menu Bar to produce histograms for the explanatory variables (refer to Chapter 1). For the Assets, use bins of width 100 starting from 0; for Sales, use bins of width 200 starting at 0; and for Profits, use bins of width 1 starting at 0. Figure 11.4 shows histograms for the three explanatory variables scores.

4. We can also examine pairwise correlations for all variables by choosing **Tools – Data Analysis – Correlation** from the Menu Bar, as in Section 2.2.

| | F | G | H | I | J | K |
|---|---|---|---|---|---|---|
| 3 | Assets | | Sales | | Profits | |
| 4 | | | | | | |
| 5 | Mean | 98.753 | Mean | 48.601 | Mean | 4.340 |
| 6 | Standard Erro | 27.149 | Standard Erro | 7.932 | Standard Erro | 0.554 |
| 7 | Median | 36.655 | Median | 28.43 | Median | 3.46 |
| 8 | Mode | #N/A | Mode | #N/A | Mode | #N/A |
| 9 | Standard Devi | 148.699 | Standard Devi | 43.443 | Standard Devi | 3.037 |
| 10 | Sample Variar | 22111.293 | Sample Variar | 1887.283 | Sample Variar | 9.221 |
| 11 | Kurtosis | 10.290 | Kurtosis | 2.125 | Kurtosis | -1.053 |
| 12 | Skewness | 3.014 | Skewness | 1.699 | Skewness | 0.529 |
| 13 | Range | 703.47 | Range | 154.11 | Range | 10.54 |
| 14 | Minimum | 13.47 | Minimum | 13.26 | Minimum | 0.18 |
| 15 | Maximum | 716.94 | Maximum | 167.37 | Maximum | 10.72 |
| 16 | Sum | 2962.6 | Sum | 1458.02 | Sum | 130.21 |
| 17 | Count | 30 | Count | 30 | Count | 30 |

Figure 11.3: Summary Statistics

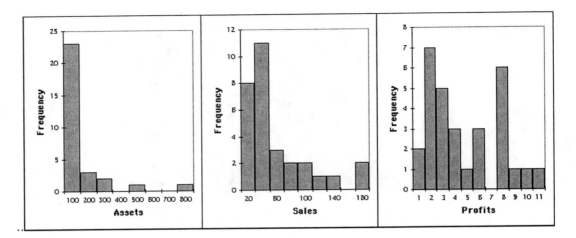

Figure 11.4: Histograms of Assets, Sales, and Profits

|   | M | N | O | P |
|---|---|---|---|---|
| 3 | | *Assets* | *Sales* | *Profits* |
| 4 | Assets | 1 | | |
| 5 | Sales | 0.4549 | 1 | |
| 6 | Profits | 0.5325 | 0.5384 | 1 |

Figure 11.5: Dow Jones Pairwise Correlations

The correlation matrix appears in Figure 11.5. All pairs of variables are moderately correlated with correlations around 0.5.

5. Finally, scatterplots may be constructed. These are shown in the text and the details for constructing scatterplots appear in Chapter 2, which may be consulted.

## Using the Regression Tool

Having explored some distributional aspects of the explanatory and response variables, we are ready to run a multiple regression.

**Example 11.2.** (Example 11.4, page 641 of text.) Use Excel's **Regression Tool** to obtain the least-squares multiple regression equation.

**Solution.** We use the same dialog box as in the one-variable regression that appears after we choose **Tools – Data Analysis – Regression** from the Menu Bar, except that we have two columns of explanatory variables for the **Input X Range** (Figure 11.6). Figure 11.7 shows a portion of the multiple regression

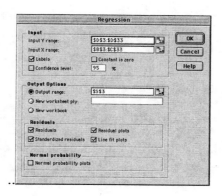

Figure 11.6: Multiple Regression Dialog Box

| | S | T | U | V | W | X | Y |
|---|---|---|---|---|---|---|---|
| 3 | SUMMARY OUTPUT | | | | | | |
| 4 | | | | | | | |
| 5 | *Regression Statistics* | | | | | | |
| 6 | Multiple R | 0.6278 | | | | | |
| 7 | R Square | 0.3942 | | | | | |
| 8 | Adjusted R Sq | 0.3493 | | | | | |
| 9 | Standard Erro | 2.4496 | | | | | |
| 10 | Observations | 30 | | | | | |
| 11 | | | | | | | |
| 12 | ANOVA | | | | | | |
| 13 | | *df* | *SS* | *MS* | *F* | *Significance F* | |
| 14 | Regression | 2 | 105.4023 | 52.7012 | 8.7829 | 0.0012 | |
| 15 | Residual | 27 | 162.0122 | 6.0005 | | | |
| 16 | Total | 29 | 267.4145 | | | | |
| 17 | | | | | | | |
| 18 | | *Coefficients* | *Standard Error* | *t Stat* | *P-value* | *Lower 95%* | *Upper 95%* |
| 19 | Intercept | 2.3405 | 0.6821 | 3.4312 | 0.0019 | 0.9409 | 3.7400 |
| 20 | Assets | 0.0074 | 0.0034 | 2.1561 | 0.0401 | 0.0004 | 0.0145 |
| 21 | Sales | 0.02610 | 0.01176 | 2.21987 | 0.03502 | 0.00198 | 0.05022 |

Figure 11.7: Multiple Regression Output

output. From cells T19:T21, we read the least-squares equation

$$\widehat{\text{Profit}} = 2.34 + 0.0074 \times \text{Assets} + 0.0261 \times \text{Sales}$$

There are other components to the output, among which are residuals and scatterplots. These will be discussed in another example in Section 11.2.

## 11.2 Inference for Multiple Regression

In the previous section, we were merely interested in descriptive statistics and in determining the least-squares regression equation for predicting profits from assets and sales. In this section, we develop a statistical model involving assumptions about the population from which the data were collected, in order to make inferences on the parameters.

As in Chapter 10, the model is

$$\text{DATA} = \text{FIT} + \text{RESIDUAL}$$

expressed mathematically as

$$y_i = \beta_0 + \beta_1 x_{i1} + \beta_2 x_{i2} + \cdots + \beta_p x_{ip} + \varepsilon_i$$

The FIT portion involving the $\{\beta_i\}$ is a linear model and expresses the population regression equation

$$\mu_y = \beta_0 + \beta_1 x_1 + \cdots + \beta_p x_p$$

which is the mean of the response variable for explanatory variables $(x_1, \ldots, x_p)$. The RESIDUAL component represents the variation in the observations, and the $\{\varepsilon_i\}$ are assumed to be independent and identically distributed $N(0, \sigma)$ random variables. The $i$th residual is $e_i = y_i - \hat{y}_i$ and is used to estimate the population variance using

$$s^2 = \frac{\sum_{i=1}^{n} e_i^2}{n - p - 1}$$

The statistics about the parameters allow you to

- construct confidence intervals of the form

$$b_j \pm t^* \text{SE}_{b_j}$$

  for the parameter $\beta_j$ where $\text{SE}_{b_j}$ is the standard error of $b_j$

- test the null hypothesis

$$H_0 : \beta_j = 0$$

  using the test statistic

$$t = \frac{b_j}{\text{SE}_{b_j}}$$

- make inferences on predicted future values and on the population means at specified values of the explanatory variables.

## Case Study

**Example 11.3.** (Case 1.2, page 652, and Examples 11.7–11.10, pages 652–661 of text. Complete data appear in the CSDATA set in the Student CD-ROM.) Data were collected at a large university on all first-year computer science majors in a particular year. The purpose of the study was to attempt to predict success in early university years. One measure of success was the cumulative grade point average (GPA) after three semesters. Among the explanatory variables recorded at the time the students enrolled in the university were average high school grades in mathematics (HSM), science (HSS), and English (HSE), coded on a scale from 1 to 10. Use the three explanatory variables, $x_1 = $ HSM, $x_2 = $ HSS, and $x_3 = $ HSE to predict the response variable, the cumulative grade point average, in a multiple regression model

$$\mu_{\text{GPA}} = \beta_0 + \beta_1 \text{ HSM} + \beta_2 \text{HSS} + \beta_3 \text{ HSE}$$

Also recorded were SATM (SAT Math score) and SATV (SAT Verbal score). The first 20 cases in the full data set (in cells A3:G227) are shown in Figure 11.8

|   | A | B | C | D | E | F | G |
|---|---|---|---|---|---|---|---|
| 1 | | | Multiple Regression Case Study | | | | |
| 2 | | | | | | | |
| 3 | Student | HSM | HSS | HSE | SATM | SATV | GPA |
| 4 | 1 | 10 | 10 | 10 | 670 | 600 | 3.32 |
| 5 | 2 | 6 | 8 | 5 | 700 | 640 | 2.26 |
| 6 | 3 | 8 | 6 | 8 | 640 | 530 | 2.35 |
| 7 | 4 | 9 | 10 | 7 | 670 | 600 | 2.08 |
| 8 | 5 | 8 | 9 | 8 | 540 | 580 | 3.38 |
| 9 | 6 | 10 | 8 | 8 | 760 | 630 | 3.29 |
| 10 | 7 | 8 | 8 | 7 | 600 | 400 | 3.21 |
| 11 | 8 | 3 | 7 | 6 | 460 | 530 | 2.00 |
| 12 | 9 | 9 | 10 | 8 | 670 | 450 | 3.18 |
| 13 | 10 | 7 | 7 | 6 | 570 | 480 | 2.34 |
| 14 | 11 | 9 | 10 | 6 | 491 | 488 | 3.08 |
| 15 | 12 | 5 | 9 | 7 | 600 | 600 | 3.34 |
| 16 | 13 | 6 | 8 | 8 | 510 | 530 | 1.40 |
| 17 | 14 | 10 | 9 | 9 | 750 | 610 | 1.43 |
| 18 | 15 | 8 | 9 | 6 | 650 | 460 | 2.48 |
| 19 | 16 | 10 | 10 | 9 | 720 | 630 | 3.73 |
| 20 | 17 | 10 | 10 | 9 | 760 | 500 | 3.8 |
| 21 | 18 | 9 | 9 | 8 | 800 | 610 | 4.00 |
| 22 | 19 | 9 | 6 | 5 | 640 | 670 | 2.00 |
| 23 | 20 | 9 | 10 | 9 | 750 | 700 | 3.74 |

Figure 11.8: GPA Data Set

## Preliminary Analysis

As in Example 11.1, it is important to use some descriptive tools as preliminary steps. Follow the steps described in Example 11.1. We have reproduced the summary statistics in Figure 11.9, histograms in Figure 11.10, and correlations in Figure 11.11.

## Estimating the Parameters

**Example 11.4** (Example 11.8, page 654. Predicting GPA from high school grades) Enter the data in consecutive columns of a worksheet as shown in Figure 11.8 and then use the **Regression Tool** as in Example 11.2. The output appears in Figure 11.12. At the bottom of this figure,

|   | H | I | J | K | L | M | N |
|---|---|---|---|---|---|---|---|
| 3 |  | HSM | HSS | HSE | SATM | SATV | GPA |
| 4 | mean | 8.3214 | 8.0893 | 8.0938 | 595.29 | 504.55 | 2.6352 |
| 5 | Std_Dev | 1.6387 | 1.6997 | 1.5079 | 86.40 | 92.61 | 0.7794 |
| 6 | minimum | 2 | 3 | 3 | 300 | 285 | 0.12 |
| 7 | maximum | 10 | 10 | 10 | 800 | 760 | 4 |

Figure 11.9: GPA Data Summary Statistics

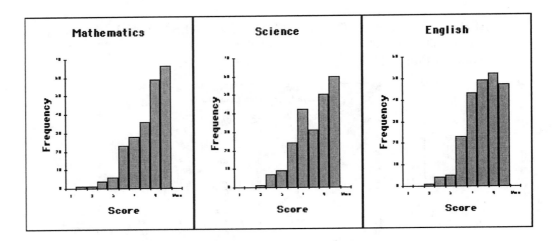

Figure 11.10: GPA Data Histograms

we find in cells T17:T20 the least-squares estimates for $\beta_0, \beta_1, \beta_2$, and $\beta_3$, respectively. Thus the fitted regression equation is

$$\widehat{\text{GPA}} = 0.590 + 0.169 \text{ HSM} + 0.034 \text{ HSS} + 0.045 \text{ HSE}.$$

Cell T7 gives the estimated standard deviation $s = 0.700$

Cells V17:V20 give $t$ statistics for testing the individual null hypotheses $H_0 : \beta_i = 0$, and cells W17:W20 give the associated $P$-values for two-sided alternatives. We see that only HSM is a significant explanatory

|   | H | I | J | K | L | M | N |
|---|---|---|---|---|---|---|---|
| 28 |  | HSM | HSS | HSE | SATM | SATV | GPA |
| 29 | HSM | 1 |  |  |  |  |  |
| 30 | HSS | 0.5757 | 1 |  |  |  |  |
| 31 | HSE | 0.4469 | 0.5794 | 1 |  |  |  |
| 32 | SATM | 0.4535 | 0.2405 | 0.1083 | 1 |  |  |
| 33 | SATV | 0.2211 | 0.2617 | 0.2437 | 0.4639 | 1 |  |
| 34 | GPA | 0.4365 | 0.3294 | 0.2890 | 0.2517 | 0.1145 | 1 |

Figure 11.11: GPA Data Pairwise Correlations

| | S | T | U | V | W | X | Y |
|---|---|---|---|---|---|---|---|
| 1 | SUMMARY OUTPUT | | | | | | |
| 2 | | | | | | | |
| 3 | *Regression Statistics* | | | | | | |
| 4 | Multiple R | 0.452 | | | | | |
| 5 | R Square | 0.205 | | | | | |
| 6 | Adjusted R Square | 0.194 | | | | | |
| 7 | Standard Error | 0.700 | | | | | |
| 8 | Observations | 224 | | | | | |
| 9 | | | | | | | |
| 10 | ANOVA | | | | | | |
| 11 | | *df* | *SS* | *MS* | *F* | *Significance F* | |
| 12 | Regression | 3 | 27.712 | 9.237 | 18.861 | 0.0000 | |
| 13 | Residual | 220 | 107.750 | 0.490 | | | |
| 14 | Total | 223 | 135.463 | | | | |
| 15 | | | | | | | |
| 16 | | *Coefficients* | *Standard Error* | *t Stat* | *P-value* | *Lower 95%* | *Upper 95%* |
| 17 | Intercept | 0.5899 | 0.2942 | 2.0047 | 0.0462 | 0.0100 | 1.1698 |
| 18 | HSM | 0.1686 | 0.0355 | 4.7494 | 0.0000 | 0.0986 | 0.2385 |
| 19 | HSS | 0.0343 | 0.0376 | 0.9136 | 0.3619 | -0.0397 | 0.1083 |
| 20 | HSE | 0.0451 | 0.0387 | 1.1655 | 0.2451 | -0.0312 | 0.1214 |

Figure 11.12: GPA Grades Regression Tool Output

variable with a *P*-value of 0.0000. In cells X17:Y20 we can find 95% confidence intervals for these coefficients.

## ANOVA Table

**Example 11.5**  (Example 11.9, page 659. *F* test for high school grades.) The ANOVA *F* statistic is given in cell W12 as 18.861. Cells T12:T13 show that the numerator degrees of freedom are DFM $= p = 3$ and the error (residual) degrees of freedom are DFE $= n - p - 1 = 224 - 3 - 1 = 220$. The *P*-value is 0.0000 in cell X12. Therefore we reject

$$H_0 : \beta_1 = \beta_2 = \beta_3 = 0$$

and conclude at least one of the three regression coefficients is not 0.

## Squared Multiple Correlation $R^2$

**Example 11.6**  (Example 11.10, page 660. $R^2$ for high school grades.) In cell T5 we see that the squared multiple correlation is

$$R^2 = 0.205$$

which indicates that 20.5% of the observed variation in the GPA scores $\{y_i\}$ is accounted for by the linear regression on these three high school scores.

**Warning.** Some of the numerical results appear contradictory. The value $R^2 = 0.205$ is small, indicating that the model does not explain much of the variation.

Yet the small $P$-value for the test $H_0 : \beta_1 = 0$ against $H_a : \beta_1 \neq 0$ suggests that the HSM score is significant. Moreover, if we ran simple regressions of GPA against each of HSS and HSE we would find that the corresponding explanatory variables taken individually are significant, while all three taken together are not.

A partial explanation for this may be found in the relatively high correlations between HSM and HSS, and between HSM and HSE, as shown in Figure 11.11. This means that there is overlap in predictive information contained in these variables.

# Chapter 12

# Statistics for Quality: Control and Capability

The goal of statistical process control is to monitor the distribution of variables, usually in a manufacturing context, and keep the observed pattern of values stable or constant over time. There will be some variation even in the most precisely produced item because of inherent changes in the raw material, adjustment of machine settings, and behavior of the operator, for instance. In this chapter we will discuss some of the tools for maintaining process control such as **control charts**.

## 12.1  Statistical Process Control

Control charts are statistical tools that monitor a process and alert us when the process has been disturbed. One of the most commonly used control charts is an $\bar{x}$ chart.

### $\bar{x}$ Chart for Process Monitoring

The preliminary steps involve the **chart setup** to establish what constitutes normal operation. Then the process is monitored.

Choose a quantitative variable $x$, which measures quality and which is assumed to have a normal distribution with mean $\mu$ and standard deviation $\sigma$. Of course, in actual practice, these values are estimated. Take small samples at regular intervals and plot the means $\bar{x}$ of these samples against time. Draw a solid line on the graph at height $\mu$. Then draw dashed lines at heights $\bar{x} \pm 3\sigma$, which mark the range of variation expected under normal operating conditions.

> **Example 12.1.** (Case 12.1, page 12.12 and Example 12.3, page 12.14 in Companion Chapter 12.) Table 12.1 on page 12.13 in Companion Chapter 12 shows 20 samples of size 4 each taken of the tensions in

fine vertical wires behind the surface of a computer monitor. The manufacturing process has been stable with mean tension $\mu = 275$ mV and process standard deviation $\sigma = 43$ mV. Construct a control chart for the sample mean.

| | A | B | C | D | E | F | G | H | I |
|---|---|---|---|---|---|---|---|---|---|
| 1 | Control Chart for the Mean | | | | | | | | |
| 2 | | | | | | | | | |
| 3 | | | | | | Sample | Center | | |
| 4 | Sample | | Tension Measurement | | | Mean | Line | Lower CL | Upper CL |
| 5 | 1 | 234.5 | 272.3 | 234.5 | 272.3 | 253.4 | 275.0 | 210.5 | 339.5 |
| 6 | 2 | 311.1 | 305.8 | 238.5 | 286.2 | 285.4 | 275.0 | 210.5 | 339.5 |
| 7 | 3 | 247.1 | 205.3 | 252.6 | 316.1 | 255.3 | 275.0 | 210.5 | 339.5 |
| 8 | 4 | 215.4 | 296.8 | 274.2 | 256.8 | 260.8 | 275.0 | 210.5 | 339.5 |
| 9 | 5 | 327.9 | 247.2 | 283.3 | 232.6 | 272.8 | 275.0 | 210.5 | 339.5 |
| 10 | 6 | 304.3 | 236.3 | 201.8 | 238.5 | 245.2 | 275.0 | 210.5 | 339.5 |
| 11 | 7 | 268.9 | 276.2 | 275.6 | 240.2 | 265.2 | 275.0 | 210.5 | 339.5 |
| 12 | 8 | 282.1 | 247.7 | 259.8 | 272.8 | 265.6 | 275.0 | 210.5 | 339.5 |
| 13 | 9 | 260.8 | 259.9 | 247.9 | 345.3 | 278.5 | 275.0 | 210.5 | 339.5 |
| 14 | 10 | 329.3 | 231.8 | 307.2 | 273.4 | 285.4 | 275.0 | 210.5 | 339.5 |
| 15 | 11 | 266.4 | 249.7 | 231.5 | 265.2 | 253.2 | 275.0 | 210.5 | 339.5 |
| 16 | 12 | 168.8 | 330.9 | 333.6 | 318.3 | 287.9 | 275.0 | 210.5 | 339.5 |
| 17 | 13 | 349.9 | 334.2 | 292.3 | 301.5 | 319.5 | 275.0 | 210.5 | 339.5 |
| 18 | 14 | 235.2 | 283.1 | 245.9 | 263.1 | 256.8 | 275.0 | 210.5 | 339.5 |
| 19 | 15 | 257.3 | 218.4 | 296.2 | 275.2 | 261.8 | 275.0 | 210.5 | 339.5 |
| 20 | 16 | 235.1 | 252.7 | 300.6 | 297.6 | 271.5 | 275.0 | 210.5 | 339.5 |
| 21 | 17 | 286.3 | 293.8 | 236.2 | 275.3 | 272.9 | 275.0 | 210.5 | 339.5 |
| 22 | 18 | 328.1 | 272.6 | 329.7 | 260.1 | 297.6 | 275.0 | 210.5 | 339.5 |
| 23 | 19 | 316.4 | 287.4 | 373.0 | 286.0 | 315.7 | 275.0 | 210.5 | 339.5 |
| 24 | 20 | 296.8 | 350.5 | 280.6 | 259.8 | 296.9 | 275.0 | 210.5 | 339.5 |

Figure 12.1: Data – $\bar{x}$ Control Chart

**Solution.**    The following steps use the **ChartWizard**. Although the sequence of steps may appear complicated, they are just an application of what was covered in Chapter 2 and involved only a few selections and actions with the worksheet.

1. Enter the data and labels from the table into an Excel worksheet as in Figure 12.1 and include additional labels as shown. The sample means of the rows have already been provided in the text. If they are not provided, then you will need to compute them yourself by entering the function = AVERAGE(B5:EF) in cell F5. Then fill down to F24 by selecting cell F5 and double-clicking the fill handle in the lower-right corner of F5. The block F5:F24 will appear selected and will be filled with the sample means of the corresponding rows.

2. Next we use the **ChartWizard** to construct a time series plot of the sample means.  Select the sample means in F5:F24 and click the ChartWizard button. In Step 1 select **Line with markers**, and in the remaining steps format the chart as shown in Figure 12.2. For instance, in Step 3 clear the **Show legend** and all **Gridlines**, and add **labels**. You may also wish to

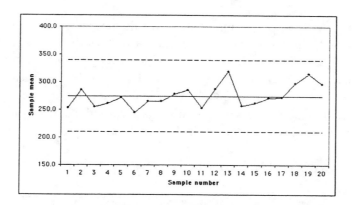

Figure 12.2: $\bar{x}$ Control Chart – $\mu$ and $\sigma$ Known

change the scale on the $y$ axis. We show how to do this in Chapter 2 and review the procedure. Double-click the $Y$ axis and in the **Format Axis** dialog box click the **Scale** tab. Uncheck the **Minimum** and **Maximum** boxes and enter the values 150 and 400, respectively. The values in the other boxes should be fine but they can be modified in the same way. Click **OK** when finished.

3. Next we enter the value for the center line in the worksheet, so enter the label **Center Line** in column G and the value 275 in cells G5:G24. There are several ways to add the center line to the time plot, and we will give all of them. The simplest way is to highlight the data (cells G5:G24) and move the pointer to the side of the selected cells until it becomes a hand and then drag it into the chart and release the mouse button. A horizontal line with markers appears at height 275. A less efficient way, but one which gives more control to the user, is to select the chart and then from the Menu Bar choose **Chart – Source Data** ... to bring up the **Source Data** dialog box. Click the **Series** tab, then click the button **Add** and in the **Values** text area enter the range G5:G24. This adds the horizontal center line. Finally, the third way includes both the sample mean and the center line data from the start. Instead of highlighting only F5:F24 prior to clicking the **ChartWizard**, highlight the data in both columns F5:G24. However you choose to include the center line, you will need to modify it to eliminate the sample markers and only show the line. This was discussed in Chapter 2. Select the center line series by clicking on one of the markers to activate the **Format Data Series** dialog box. If the **Patterns** tab is not selected, then click it and change the **Marker** button to None. You can also repeat this process for the series of sample means to change the type of marker.

4. Finally, we add the lower and upper control limits. These require the standard

deviation. We use the assumed standard deviation $\sigma = 43$ mV, although in practice we might use the sample standard deviation. Enter the value 43 in cell H5 and fill to cell H24. In cell I5 enter the lower limit $\mu - 3\frac{43}{\sqrt{4}} = 275 - 64.5 = 210.5$ mV fill to I24, and in cell J4 enter the upper limit $\mu + 3\frac{43}{\sqrt{4}} = 275 + 64.5 = 339.5$ mV. Finally add lower and upper lines to the chart in exactly the same fashion that you added the center line. After formatting, final result appears in Figure 12.2.

## $s$ Chart for Process Monitoring

| | A | B | C | D | E | F | G | H | I |
|---|---|---|---|---|---|---|---|---|---|
| 1 | Control Chart for the Standard Deviation | | | | | | | | |
| 2 | | | | | | | | | |
| 3 | | | | | | Sample | Center | | |
| 4 | Sample | | Tension Measurement | | | StDev | Line | Lower CL | Upper CL |
| 5 | 1 | 234.5 | 272.3 | 234.5 | 272.3 | 21.8 | 39.6 | 0 | 89.8 |
| 6 | 2 | 311.1 | 305.8 | 238.5 | 286.2 | 33.0 | 39.6 | 0 | 89.8 |
| 7 | 3 | 247.1 | 205.3 | 252.6 | 316.1 | 45.7 | 39.6 | 0 | 89.8 |
| 8 | 4 | 215.4 | 296.8 | 274.2 | 256.8 | 34.4 | 39.6 | 0 | 89.8 |
| 9 | 5 | 327.9 | 247.2 | 283.3 | 232.6 | 42.5 | 39.6 | 0 | 89.8 |
| 10 | 6 | 304.3 | 236.3 | 201.8 | 238.5 | 42.8 | 39.6 | 0 | 89.8 |
| 11 | 7 | 268.9 | 276.2 | 275.6 | 240.2 | 17.0 | 39.6 | 0 | 89.8 |
| 12 | 8 | 282.1 | 247.7 | 259.8 | 272.8 | 15.0 | 39.6 | 0 | 89.8 |
| 13 | 9 | 260.8 | 259.9 | 247.9 | 345.3 | 44.9 | 39.6 | 0 | 89.8 |
| 14 | 10 | 329.3 | 231.8 | 307.2 | 273.4 | 42.5 | 39.6 | 0 | 89.8 |
| 15 | 11 | 266.4 | 249.7 | 231.5 | 265.2 | 16.3 | 39.6 | 0 | 89.8 |
| 16 | 12 | 168.8 | 330.9 | 333.6 | 318.3 | 79.7 | 39.6 | 0 | 89.8 |
| 17 | 13 | 349.9 | 334.2 | 292.3 | 301.5 | 27.1 | 39.6 | 0 | 89.8 |
| 18 | 14 | 235.2 | 283.1 | 245.9 | 263.1 | 21.0 | 39.6 | 0 | 89.8 |
| 19 | 15 | 257.3 | 218.4 | 296.2 | 275.2 | 33.0 | 39.6 | 0 | 89.8 |
| 20 | 16 | 235.1 | 252.7 | 300.6 | 297.6 | 32.7 | 39.6 | 0 | 89.8 |
| 21 | 17 | 286.3 | 293.8 | 236.2 | 275.3 | 25.6 | 39.6 | 0 | 89.8 |
| 22 | 18 | 328.1 | 272.6 | 329.7 | 260.1 | 36.5 | 39.6 | 0 | 89.8 |
| 23 | 19 | 316.4 | 287.4 | 373.0 | 286.0 | 40.7 | 39.6 | 0 | 89.8 |
| 24 | 20 | 296.8 | 350.5 | 280.6 | 259.8 | 38.8 | 39.6 | 0 | 89.8 |

Figure 12.3: Data – $s$ Control Chart

Next we describe a control chart for monitoring the standard deviation. Unlike the sample mean $\bar{x}$, the sample standard deviation $s$ does not have a normal distribution. Nonetheless, for the practical applications of control charts, it is still customary to use the $3\sigma$ rule and draw a center line on the chart at the height $\mu_s$ and lower and upper control limits at three standard deviation units below or above the center line, that is at

$$\text{LCL} = \mu_s - 3\sigma_s = (c_4 - 3c_5)\sigma = B_5\sigma$$
$$\text{UCL} = \mu_s + 3\sigma_s = (c_4 + 3c_5)\sigma = B_6\sigma$$

where the values $c_4, c_5, B_5$, and $B_6$ have been tablulated for values $2 \leq n \leq 10$, $n$ being the sample size of the averages plotted, in Table 12.3, page 12.19 in Companion Chapter 12.

**Example 12.2.**    (Example 12.4, page 12.19 in Companion Chapter 12.) Construct an $s$ chart for the data in Example 12.1.

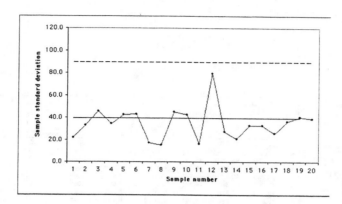

Figure 12.4: $s$ Control Chart – $\mu$ and $\sigma$ Known

**Solution.**    The center line is at $c_4\sigma = (0.9213)(43) = 39.6$ mV and

$$\text{LCL} = B_5\sigma = (0)(43) = 0$$
$$\text{UCL} = B_6\sigma = (2.088)(43) = 89.8$$

Now repeat the steps in Example 12.1 modifying the worksheet by replacing the sample mean data with the sample standard deviations (Figure 12.3) to arrive at the control chart in Figure 12.4.

## 12.2   Using Control Charts

In practice $\mu$ and $\sigma$ may not be known, especially when setting up a control chart at the start of production. Data need to be collected to estimate $\mu$ and $\sigma$. What we then do is modify the $\bar{x}$ chart and the $s$ chart by taking regular samples of size $n$ from the process and then use estimates

$$\hat{\mu} = \bar{\bar{x}} \quad \text{in place of } \mu$$
$$\hat{\sigma} = \frac{\bar{s}}{c_4} \quad \text{in place of } \sigma$$

where $\bar{s}$ is the average of the sample standard deviations and the constant $c_4$, from Table 12.3 referenced above, corrects for the bias of $s$ in small samples. Thus
  • for an $\bar{x}$ chart

$$\text{LCL} = \hat{\mu} - 3\frac{\hat{\sigma}}{\sqrt{n}}$$
$$\text{CL} = \hat{\mu}$$
$$\text{UCL} = \hat{\mu} + 3\frac{\hat{\sigma}}{\sqrt{n}}$$

• for an $s$ chart

$$LCL = B_5 \hat{\sigma}$$
$$CL = c_4 \hat{\sigma} = \bar{s}$$
$$UCL = B_6 \hat{\sigma}$$

| | A | B | C | D | E | F | G | H | I |
|---|---|---|---|---|---|---|---|---|---|
| 1 | Setting up a Control Chart | | | | | | | | |
| 2 | | | | | | | | | |
| 3 | | Sample | Center | xbar | xbar | Sample | Center | s | s |
| 4 | Sample | Mean | Line | Lower CL | Upper CL | StDev | Line | Lower CL | Upper CL |
| 5 | 1 | 49.750 | 48.380 | 46.741 | 50.019 | 2.684 | 1.0065 | 0 | 2.281 |
| 6 | 2 | 49.375 | 48.380 | 46.741 | 50.019 | 0.895 | 1.0065 | 0 | 2.281 |
| 7 | 3 | 50.250 | 48.380 | 46.741 | 50.019 | 0.895 | 1.0065 | 0 | 2.281 |
| 8 | 4 | 49.875 | 48.380 | 46.741 | 50.019 | 1.118 | 1.0065 | 0 | 2.281 |
| 9 | 5 | 47.250 | 48.380 | 46.741 | 50.019 | 0.671 | 1.0065 | 0 | 2.281 |
| 10 | 6 | 45.000 | 48.380 | 46.741 | 50.019 | 2.684 | 1.0065 | 0 | 2.281 |
| 11 | 7 | 48.375 | 48.380 | 46.741 | 50.019 | 0.671 | 1.0065 | 0 | 2.281 |
| 12 | 8 | 48.500 | 48.380 | 46.741 | 50.019 | 0.447 | 1.0065 | 0 | 2.281 |
| 13 | 9 | 48.500 | 48.380 | 46.741 | 50.019 | 0.447 | 1.0065 | 0 | 2.281 |
| 14 | 10 | 46.250 | 48.380 | 46.741 | 50.019 | 1.566 | 1.0065 | 0 | 2.281 |
| 15 | 11 | 49.000 | 48.380 | 46.741 | 50.019 | 0.895 | 1.0065 | 0 | 2.281 |
| 16 | 12 | 48.125 | 48.380 | 46.741 | 50.019 | 0.671 | 1.0065 | 0 | 2.281 |
| 17 | 13 | 47.875 | 48.380 | 46.741 | 50.019 | 1.118 | 1.0065 | 0 | 2.281 |
| 18 | 14 | 48.250 | 48.380 | 46.741 | 50.019 | 0.895 | 1.0065 | 0 | 2.281 |
| 19 | 15 | 47.625 | 48.380 | 46.741 | 50.019 | 0.671 | 1.0065 | 0 | 2.281 |
| 20 | 16 | 47.375 | 48.380 | 46.741 | 50.019 | 0.671 | 1.0065 | 0 | 2.281 |
| 21 | 17 | 50.250 | 48.380 | 46.741 | 50.019 | 1.566 | 1.0065 | 0 | 2.281 |
| 22 | 18 | 47.000 | 48.380 | 46.741 | 50.019 | 0.895 | 1.0065 | 0 | 2.281 |
| 23 | 19 | 47.000 | 48.380 | 46.741 | 50.019 | 0.447 | 1.0065 | 0 | 2.281 |
| 24 | 20 | 49.625 | 48.380 | 46.741 | 50.019 | 1.118 | 1.0065 | 0 | 2.281 |
| 25 | 21 | 49.875 | 48.380 | 46.741 | 50.019 | 0.447 | 1.0065 | 0 | 2.281 |
| 26 | 22 | 47.625 | 48.380 | 46.741 | 50.019 | 1.118 | 1.0065 | 0 | 2.281 |
| 27 | 23 | 49.750 | 48.380 | 46.741 | 50.019 | 0.671 | 1.0065 | 0 | 2.281 |
| 28 | 24 | 48.625 | 48.380 | 46.741 | 50.019 | 0.895 | 1.0065 | 0 | 2.281 |
| 29 | | | | | | | | | |
| 30 | average xbar | 48.380 | | | average s | 1.0065 | | | |
| 31 | | | | | sigma hat | 1.0925 | | | |

Figure 12.5: Data – $\bar{x}$ Control Chart

**Example 12.3.**   (Case 12.3, pages 12.30–12.35 and in Companion Chapter 12.) The viscosity of a material is its resistance to flow when under stress. A specialty chemical company is beginning production of an elastomer that is supposed to have viscosity $45 \pm 5$ Mooneys. Table 12.5 on page 12.31 in Companion Chapter 12 shows $\bar{x}$ and $s$ from samples of size $n = 4$ lots from the first 24 shifts as production begins. Construct an $\bar{x}$ and an $s$ chart.

**Solution.**   Enter the data into columns of an Excel worksheet as shown in Figure 12.5. Make sure that you leave three columns between the sample means and the sample standard deviations, and a row at the bottom for computing the estimates.

1. Compute the estimated average $\bar{\bar{x}}$ by entering the formula = AVERAGE(B5:B28) where B5:B28 is the range for the sample means in the worksheet. (Your range may be different depending on where you choose to enter the data.) We find that $\bar{\bar{x}} = 48.380$   (in cell B30). Transfer this value to the cells C5:C28

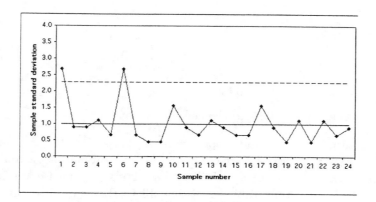

Figure 12.6: $s$ Control Chart – Two Values Out of Control

in the column labelled **Center Line**, which will be used for drawing the center line on the control chart. Similarly compute the averages $\bar{s}$ of the sample standard deviations by entering the formula = `STDEV(F5:F28)`, where F5:F28 is the range for the sample standard deviations in the worksheet. Then divide by $c_4 = 0.9213$ to get $\hat{\sigma}$. We obtain

| | A | B | C | D | E | F | G | H | I |
|---|---|---|---|---|---|---|---|---|---|
| 1 | Removing Out of Control Samples | | | | | | | | |
| 2 | | | | | | | | | |
| 3 | | Sample | Center | xbar | xbar | Sample | Center | s | s |
| 4 | Sample | Mean | Line | Lower CL | Upper CL | StDev | Line | Lower CL | Upper CL |
| 5 | 1 | | 48.472 | 47.081 | 49.862 | | 0.8540 | 0 | 1.935 |
| 6 | 2 | 49.375 | 48.472 | 47.081 | 49.862 | 0.895 | 0.8540 | 0 | 1.935 |
| 7 | 3 | 50.250 | 48.472 | 47.081 | 49.862 | 0.895 | 0.8540 | 0 | 1.935 |
| 8 | 4 | 49.875 | 48.472 | 47.081 | 49.862 | 1.118 | 0.8540 | 0 | 1.935 |
| 9 | 5 | 47.250 | 48.472 | 47.081 | 49.862 | 0.671 | 0.8540 | 0 | 1.935 |
| 10 | 6 | | 48.472 | 47.081 | 49.862 | | 0.8540 | 0 | 1.935 |
| 11 | 7 | 48.375 | 48.472 | 47.081 | 49.862 | 0.671 | 0.8540 | 0 | 1.935 |
| 12 | 8 | 48.500 | 48.472 | 47.081 | 49.862 | 0.447 | 0.8540 | 0 | 1.935 |
| 13 | 9 | 48.500 | 48.472 | 47.081 | 49.862 | 0.447 | 0.8540 | 0 | 1.935 |
| 14 | 10 | 46.250 | 48.472 | 47.081 | 49.862 | 1.566 | 0.8540 | 0 | 1.935 |
| 15 | 11 | 49.000 | 48.472 | 47.081 | 49.862 | 0.895 | 0.8540 | 0 | 1.935 |
| 16 | 12 | 48.125 | 48.472 | 47.081 | 49.862 | 0.671 | 0.8540 | 0 | 1.935 |
| 17 | 13 | 47.875 | 48.472 | 47.081 | 49.862 | 1.118 | 0.8540 | 0 | 1.935 |
| 18 | 14 | 48.250 | 48.472 | 47.081 | 49.862 | 0.895 | 0.8540 | 0 | 1.935 |
| 19 | 15 | 47.625 | 48.472 | 47.081 | 49.862 | 0.671 | 0.8540 | 0 | 1.935 |
| 20 | 16 | 47.375 | 48.472 | 47.081 | 49.862 | 0.671 | 0.8540 | 0 | 1.935 |
| 21 | 17 | 50.250 | 48.472 | 47.081 | 49.862 | 1.566 | 0.8540 | 0 | 1.935 |
| 22 | 18 | 47.000 | 48.472 | 47.081 | 49.862 | 0.895 | 0.8540 | 0 | 1.935 |
| 23 | 19 | 47.000 | 48.472 | 47.081 | 49.862 | 0.447 | 0.8540 | 0 | 1.935 |
| 24 | 20 | 49.625 | 48.472 | 47.081 | 49.862 | 1.118 | 0.8540 | 0 | 1.935 |
| 25 | 21 | 49.875 | 48.472 | 47.081 | 49.862 | 0.447 | 0.8540 | 0 | 1.935 |
| 26 | 22 | 47.625 | 48.472 | 47.081 | 49.862 | 1.118 | 0.8540 | 0 | 1.935 |
| 27 | 23 | 49.750 | 48.472 | 47.081 | 49.862 | 0.671 | 0.8540 | 0 | 1.935 |
| 28 | 24 | 48.625 | 48.472 | 47.081 | 49.862 | 0.895 | 0.8540 | 0 | 1.935 |
| 29 | | | | | | | | | |
| 30 | average xbar | 48.472 | | | average s | 0.8540 | | | |
| 31 | | | | | sigma hat | 0.9270 | | | |

Figure 12.7: Data – $\bar{x}$ Control Chart – Two Values Out of Control

$$\bar{s} \;=\; 1.0065 \quad \text{(in cell E30)}$$

$$\hat{\sigma} \;=\; \frac{\bar{s}}{0.9213} = 1.0925 \quad \text{(in cell F31)}$$

Transfer the value 1.0065 to the cells G5:G28.

2. Next determine lower and upper control limts. For the $\bar{x}$ chart enter = B\$30 - F\$31*3/SQRT(4) in cell D5 and fill to D28, and also enter = B\$30 + F\$31*3/SQRT(4) in cell E5 and fill to E28. For the $s$ chart enter 0 in cell H5 (because $B_5 = 0$ from Table 12.3) and fill to H28, and also enter = 2.088 *F\$31 in cell I5 (because $B_6 = 2.088$ from Table 12.3) and fill to I28.

3. **Step 1.** Construct an $s$ chart exactly as in Example 12.1.

4. **Step 2.** From Figure 12.6 we see that samples 1 and 6 are outside the control limits. Remove them (merely delete the standard deviations for these samples from the worksheet and all values will be recalculated automatically) and recalculate based on 22 samples that $\bar{s} = 0.8540$ and $\hat{\sigma} = 0.9270$. Now recalculate $\bar{x}$ and the associated control limits with these values. Again, the simplest way is to remove them from your worksheet.

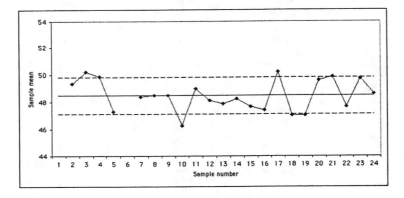

Figure 12.8: $\bar{x}$ Control Chart

5. **Step 3.** Finally, with the remaining 22 samples, make an $\bar{x}$ chart. However, in order to keep the center line and the control limit, you still need to include these corresponding limits for all 24 sample means. The modified worksheet is shown in Figure 12.7, and the corresponding $\bar{x}$ chart is shown in Figure 12.8.

# Chapter 13

# Time Series Forecasting

A **Time Series** is a sequence of measurements of a variable taken at regular intervals over time. Typically, adjacent observations are usually close; observations far apart are not. Many economic data such as stock prices or sales figures show such a behavior and plots of the data indicate trends and seasonal patterns, in addition to random fluctuations. The first step is to plot the data. Then techniques from regression analysis can be used to carry out analyses.

Table 13.1: Quarterly Retail Sales for JC Penney 1996–2001 (millions of dollars)

| Year-Quarter | Sales | Year-Quarter | Sales |
|---|---|---|---|
| 1996-1st | 4452 | 1999-1st | 7339 |
| 1996-2nd | 4507 | 1999-2nd | 7104 |
| 1996-3rd | 5537 | 1999-3rd | 7639 |
| 1996-4th | 8157 | 1999-4th | 9661 |
| 1997-1st | 6481 | 2000-1st | 7528 |
| 1997-2nd | 6420 | 2000-2nd | 7207 |
| 1997-3rd | 7208 | 2000-3rd | 7538 |
| 1997-4th | 9509 | 2000-4th | 9673 |
| 1998-1st | 6755 | 2001-1st | 7522 |
| 1998-2nd | 6483 | 2001-2nd | 7211 |
| 1998-3rd | 7129 | 2001-3rd | 7729 |
| 1998-4th | 9072 | 2001-4th | 9442 |

## 13.1   Trends and Seasons

**Example 13.1.**   (Exercise 13.1, page 13.2 in Companion Chapter 13.)
Table 13.1 contains retail sales for JC Penney in millions of dollars.

(a) Make a time plot of the data.

| | A | B | C |
|---|---|---|---|
| 1 | JC Penney Quarterly Retail Sales | | |
| 2 | | | |
| 3 | Year-Quarter | Sales | |
| 4 | 1996-1st | 4452 | |
| 5 | 1996-2nd | 4507 | |
| 6 | 1996-3rd | 5537 | |
| 7 | 1996-4th | 8157 | |
| 8 | 1997-1st | 6481 | |
| 9 | 1997-2nd | 6420 | |
| 10 | 1997-3rd | 7208 | |
| 11 | 1997-4th | 9509 | |
| 12 | 1998-1st | 6755 | |
| 13 | 1998-2nd | 6483 | |
| 14 | 1998-3rd | 7129 | |
| 15 | 1998-4th | 9072 | |
| 16 | 1999-1st | 7339 | |
| 17 | 1999-2nd | 7104 | |
| 18 | 1999-3rd | 7639 | |
| 19 | 1999-4th | 9661 | |
| 20 | 2000-1st | 7528 | |
| 21 | 2000-2nd | 7207 | |
| 22 | 2000-3rd | 7538 | |
| 23 | 2000-4th | 9673 | |
| 24 | 2001-1st | 7522 | |
| 25 | 2001-2nd | 7211 | |
| 26 | 2001-3rd | 7729 | |
| 27 | 2001-4th | 9442 | |

Figure 13.1: JC Penney Quarterly Retail Sales Data

(b) Is there an obvious trend? If so, is the trend positive or negative?

(c) Is there an obvious repeating pattern in the data? If so, describe the repeating pattern.

**Solution.**

(a) Enter the data into two columns of an Excel worksheet as in Figure 13.1. Then use the **ChartWizard** to construct a timeplot as an Excel Scatterplot shown in Figure 13.2. Click one of the points to bring up the **Format Data Series** dialog box and check the radio button **Automatic** under **Line** to connect the points by a line.

(b) There is an overall positive (upward) trend.

(c) There is a repeating yearly pattern with a small drop in each second quarter and a large jump in each fourth quarter.

## Trends

The JC Penney data can be used to forecast or predict future sales. As a first step we can add a trendline to the time series plot.

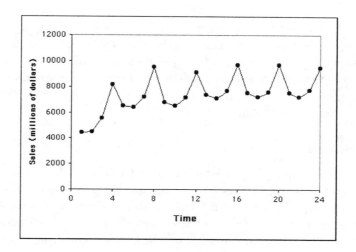

Figure 13.2: Time Plot–JC Penney Quarterly Retail Sales

**Example 13.2.** (Exercise 13.3, page 13.5 in Companion Chapter 13.)
Find the least-squares line for the JC Penney data.

**Solution.** Select one of the points in the time plot. Then from the Menu Bar choose **Chart − Add Trendline ...** and under the **Type** tab, highlight the selection **Linear**. Under the **Options** tab, check the box to display the equation

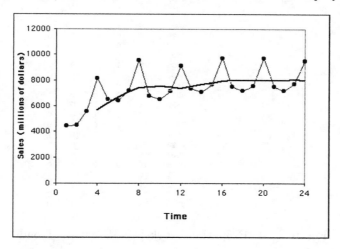

Figure 13.3: 4-Point Moving Average–JC Penney Quarterly Retail Sales

on the chart. Click **OK**. The least-squares regression line is

$$\widehat{\text{SALES}} = 5907.6 + 118.4 \times \text{year}$$

**Note:** Instead of using a linear regression line, we may wish to use a 4-point moving average. This is also a choice under the **Type** tab. A moving average takes more account of seasonal variation and makes the trend more apparent. For instance, if we were to use a 4-point moving average here, we would obtain a curve that is tapering off rather than maintaining a steady slope. This is more consistent with the data. See Figure 13.3.

## Seasonal Patterns

To improve the accuracy of forecasts, we need to account for seasonal variation in the time series. In Excel this is achieved by using categorical or **indicator variables** in **multiple regression**. The number of indicator variables included in the multiple regression model is one less than the number of categories because the effect of the remaining variables is absorbed by the constant term.

> **Example 13.3.** (Exercise 13.5, page 13.7 in Companion Chapter 13.) Sales seem to follow a pattern of ups and downs that repeats every four quarters. Add indicator variables to the linear trend model fit in Example 13.2 and compare the two models.

| | A | B | C | D | E | F | G | H |
|---|---|---|---|---|---|---|---|---|
| 1 | JC Penney Quarterly Retail Sales | | | | | | | |
| 2 | | | | | | | | |
| 3 | Year | Quarter | Sales | Time | Q1 | Q2 | Q3 | Q4 |
| 4 | 1996 | 1 | 4452 | 1 | 1 | 0 | 0 | 0 |
| 5 | | 2 | 4507 | 2 | 0 | 1 | 0 | 0 |
| 6 | | 3 | 5537 | 3 | 0 | 0 | 1 | 0 |
| 7 | | 4 | 8157 | 4 | 0 | 0 | 0 | 1 |
| 8 | 1997 | 1 | 6481 | 5 | 1 | 0 | 0 | 0 |
| 9 | | 2 | 6420 | 6 | 0 | 1 | 0 | 0 |
| 10 | | 3 | 7208 | 7 | 0 | 0 | 1 | 0 |
| 11 | | 4 | 9509 | 8 | 0 | 0 | 0 | 1 |
| 12 | 1998 | 1 | 6755 | 9 | 1 | 0 | 0 | 0 |
| 13 | | 2 | 6483 | 10 | 0 | 1 | 0 | 0 |
| 14 | | 3 | 7129 | 11 | 0 | 0 | 1 | 0 |
| 15 | | 4 | 9072 | 12 | 0 | 0 | 0 | 1 |
| 16 | 1999 | 1 | 7339 | 13 | 1 | 0 | 0 | 0 |
| 17 | | 2 | 7104 | 14 | 0 | 1 | 0 | 0 |
| 18 | | 3 | 7639 | 15 | 0 | 0 | 1 | 0 |
| 19 | | 4 | 9661 | 16 | 0 | 0 | 0 | 1 |
| 20 | 2000 | 1 | 7528 | 17 | 1 | 0 | 0 | 0 |
| 21 | | 2 | 7207 | 18 | 0 | 1 | 0 | 0 |
| 22 | | 3 | 7538 | 19 | 0 | 0 | 1 | 0 |
| 23 | | 4 | 9673 | 20 | 0 | 0 | 0 | 1 |
| 24 | 2001 | 1 | 7522 | 21 | 1 | 0 | 0 | 0 |
| 25 | | 2 | 7211 | 22 | 0 | 1 | 0 | 0 |
| 26 | | 3 | 7729 | 23 | 0 | 0 | 1 | 0 |

Figure 13.4: Preparing Worksheet–Seasonal Trend

**Solution.** In this case, the categorical variables are the quarters, so we use three indicator variables. We let

$$X1 = \begin{cases} 1 & \text{for the 1st quarter} \\ 0 & \text{otherwise} \end{cases}$$

The output, without graphs, is shown in Figure 13.6

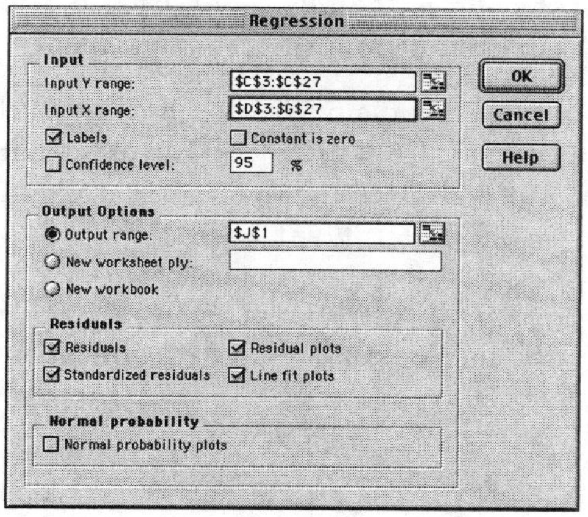


Figure 13.5: Multiple Regression Dialog Box – Seasonal Trend

| | J | K | L | M | N | O | P |
|---|---|---|---|---|---|---|---|
| 1 | SUMMARY OUTPUT | | | | | | |
| 2 | | | | | | | |
| 3 | *Regression Statistics* | | | | | | |
| 4 | Multiple R | 0.93061672 | | | | | |
| 5 | R Square | 0.86604749 | | | | | |
| 6 | Adjusted R Sq | 0.83784696 | | | | | |
| 7 | Standard Erro | 571.606245 | | | | | |
| 8 | Observations | 24 | | | | | |
| 9 | | | | | | | |
| 10 | ANOVA | | | | | | |
| 11 | | *df* | *SS* | *MS* | *F* | *Significance F* | |
| 12 | Regression | 4 | 40136395.3 | 10034098.8 | 30.7103273 | 4.6894E-08 | |
| 13 | Residual | 19 | 6207940.28 | 326733.699 | | | |
| 14 | Total | 23 | 46344335.6 | | | | |
| 15 | | | | | | | |
| 16 | | *Coefficients* | *Standard Error* | *t Stat* | *P-value* | *Lower 95%* | *Upper 95%* |
| 17 | Intercept | 7863.75833 | 334.116767 | 23.5359584 | 1.6202E-15 | 7164.44369 | 8563.07298 |
| 18 | Time | 99.1839286 | 17.0800034 | 5.80702042 | 1.3557E-05 | 63.4350595 | 134.932798 |
| 19 | Q1 | -2275.2815 | 333.971214 | -6.8128074 | 1.6692E-06 | -2974.2915 | -1576.2715 |
| 20 | Q2 | -2565.2988 | 331.780257 | -7.7319212 | 2.77E-07 | -3259.7231 | -1870.8745 |
| 21 | Q3 | -2023.1494 | 330.458711 | -6.1222457 | 6.9335E-06 | -2714.8077 | -1331.4912 |

Figure 13.6: Multiple Regression Output

$$X2 = \begin{cases} 1 & \text{for the 2nd quarter} \\ 0 & \text{otherwise} \end{cases}$$

$$X3 = \begin{cases} 1 & \text{for the 3rd quarter} \\ 0 & \text{otherwise} \end{cases}$$

The multiple regression model is

$$\text{SALES} = \beta_0 + \beta_1 \text{year} + \beta_2 X1 + \beta_2 X2 + \beta_3 X3$$

1. Prepare the worksheet by including the "0,1" values for $X1, X2, X3$, as shown in Figure 13.4. From the Menu Bar, choose **Tools − Data Analysis ...** and select **Regression** from the choices given.

2. Complete the **Regression** dialog box as in Figure 13.5 using C3:C27 for the **Input Y range** and D3:G27 as the **Input X range**. Remember only to include 3 indicator variables for the seasons. Check the box for **Labels**. Insert a location in the **Output range** text box and

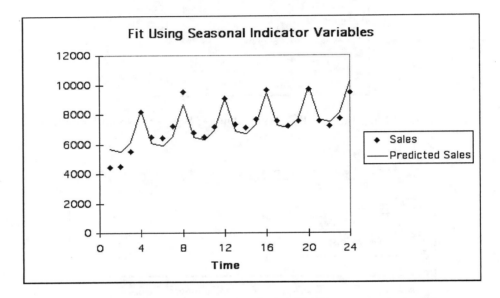

Figure 13.7: Multiple Regression Seasonal Fit

The output from the Coefficients section of Figure 13.6 gives the seasonally fitted equation as

$$\widehat{\text{SALES}} = 7863.76 + 99.18\text{year} - 2275.28X1 - 2565.30X2 - 2023.15X3$$

The output also shows the fitted values plotted against the data. We can edit the fitted values by including lines joining adjacent points to judge how well the seasonal model fits. This is shown in Figure 13.7.

We can also compare the coefficient of determination $r^2$. For the ordinary least-squares line it is 0.348. With the seasonal variables, it increases to 0.866 (cell K5 in Figure 13.6).

## Autocorrelation

Included in the output of the **Regression Tool** are residual plots, among which is the plot of residuals against time. This is shown in Figure 13.8. This plot is a

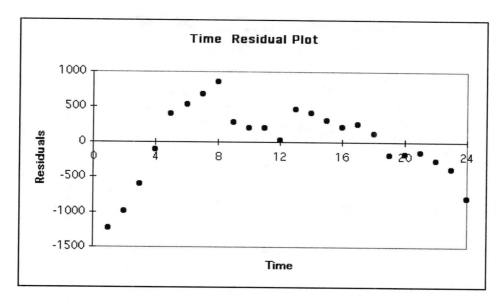

Figure 13.8: Residuals from Seasonal Fit

typical illustration of positive correlated residuals. The errors tend to be of the same sign.

Although the coefficient of determination of 0.866 is considered large, that does not mean that the model fit is good. In particular, it is part of the theory of least squares that the estimates only have good properties if the errors are independent and identically distributed. Because the residuals mimic the errors, any perceived dependence between residuals indicates that the underlying model for the errors needs to be changed.

## Lagged Residual Plot

The correlation between successive values of the dependent variable $y_{t-1}$ and $y_t$ is called the auto-correlation.

> **Example 13.4.** Construct a lagged plot of the JC Penney data.

**Solution.**

1. In the original worksheet insert a column labelled "Lag" adjacent to the sales data. In our worksheet (Figure 13.9) we have moved the sales data to the right by one column and inserted the Lag column to the left. Copy the first 23 observations of the sales data (all except the last value) and paste it into cells C5:C27. Thus column C contains the same data as column D but is "lagged" behind by one time point.

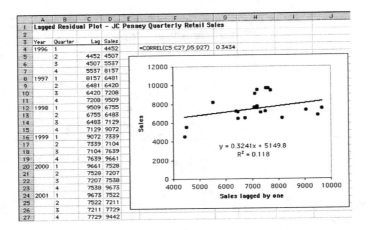

Figure 13.9: Lagged Residual Plot

2. Using the **ChartWizard** construct a scatterplot (Figure 13.9) of the data in cells C5:D27. The block contains the pairs of points $y_t, y_{t-1}$ for $2 \leq t \leq 24$.

3. The residuals show a linear trend, so add a trendline and include the equation and $r^2$ on the plot. We have also computed the correlation coefficient 0.3434 in cell G4, using the formula shown in cell F4.

## Durbin-Watson Statistic

| | J | K | L | M |
|---|---|---|---|---|
| 23 | Durbin-Watson Statistic | | | |
| 24 | | | | |
| 25 | RESIDUAL OUTPUT | | | |
| 26 | | | | |
| 27 | Observation | Predicted Sales | Residuals | |
| 28 | 1 | 5687.661 | -1235.661 | |
| 29 | 2 | 5496.827 | -989.827 | |
| 30 | 3 | 6138.161 | -601.161 | |
| 31 | 4 | 8260.494 | -103.494 | |
| 32 | 5 | 6084.396 | 396.604 | |
| 33 | 6 | 5893.563 | 526.437 | |
| 34 | 7 | 6534.896 | 673.104 | |
| 35 | 8 | 8657.230 | 851.770 | |
| 36 | 9 | 6481.132 | 273.868 | |
| 37 | 10 | 6290.299 | 192.701 | |
| 38 | 11 | 6931.632 | 197.368 | |
| 39 | 12 | 9053.965 | 18.035 | |
| 40 | 13 | 6877.868 | 461.132 | |
| 41 | 14 | 6687.035 | 416.965 | |
| 42 | 15 | 7328.368 | 310.632 | |
| 43 | 16 | 9450.701 | 210.299 | |
| 44 | 17 | 7274.604 | 253.396 | |
| 45 | 18 | 7083.770 | 123.230 | |
| 46 | 19 | 7725.104 | -187.104 | |
| 47 | 20 | 9847.437 | -174.437 | |
| 48 | 21 | 7671.339 | -149.339 | |
| 49 | 22 | 7480.506 | -269.506 | |
| 50 | 23 | 8121.839 | -392.839 | |
| 51 | 24 | 10244.173 | -802.173 | |
| 52 | | | | |
| 53 | | | | |
| 54 | =SUMXMY2(L29:L51,L28:L50)/SUMSQ(L28:L50) | | | 0.3028 |

Figure 13.10: Durbin-Watson Statistic

There is a formal procedure for testing for autocorrelation of residuals in a time series, called the Durbin-Watson test. The **Regression Tool** provides residuals in the "Residual Output" section of the output, shown in Figure 13.10. From this we can easily compute the DW statistic. It is given by

$$\text{Durbin-Watson statistic} = \frac{\sum_{i=2}^{n}(y_i - y_{i-1})^2}{\sum_{i=2}^{n} y_i^2}$$

This formula is easily translated into Excel by

$$= \text{SUMXMY2}(\text{L29}:\text{L51}, \text{L28}:\text{L50})/\text{SUMSQ}(\text{L28}:\text{L50})$$

and equals 0.3018. Values of the DW statistic close to 0 are an indicator of strong positive auto-correlation.

# Chapter 14

# One-Way Analysis of Variance

In Chapter 7 we were interested in comparing the means of two populations. For the case of independent samples from each population, the statistical distribution used was Student's $t$ statistics. In this chapter we will be comparing the means of $I \geq 2$ populations.

We might approach the analysis by taking each of the $I$ populations in pairs. For moderate $I$, this isn't a bad idea as long as Bonferroni adjustments are made to the significance levels or the confidence levels to account for multiple comparisons – see Example 14.23, page 14.36 in Companion Chapter 14 – or simultaneous inference procedures are used. These provide comparisons between pairs of population means or contrasts. However, there is a simpler procedure for testing whether all population means are the same that reduces to the square of the two-sample $t$ when $I = 2$. This is called analysis of variance. In particular, the $F$ statistic in ANOVA, when there are two populations, is precisely the square of Student's $t$, and the ANOVA $F$ test is then identical to the Student two-sample $t$ procedure.

As the name suggests, ANOVA consists of separating the variability in a data set into two components and judging whether a fit that assumes equal population means is substantially better than a fit in which all means are assumed to be the same. This is achieved by comparing the residual variation following the model fit in both cases by a ratio called an $F$, which may be viewed as a signal-to-noise ratio. (See also the discussion in Chapter 10 on the ANOVA output from the **Regression Tool**.)

## 14.1 One-Way Analysis of Variance

### An Overview of ANOVA

Suppose there are $I$ normal populations labeled $1 \leq i \leq I$ and that independent random samples $\{x_{ij} : 1 \leq j \leq n_i\}$ of size $n_i \geq 1$, $1 \leq i \leq I$ are taken from each.

The ANOVA model is

$$x_{ij} = \mu_i + \varepsilon_{ij} \qquad 1 \le i \le I,\ 1 \le j \le n_i$$

where $\{\varepsilon_{ij}\}$ are independent $N(0, \sigma)$ random variables. The populations thus have means $\mu_i$ and a common standard deviation $\sigma$. We express this as

$$\text{DATA} = \text{FIT} + \text{RESIDUAL}$$

The one-way ANOVA significance test is

$$H_0 : \mu_1 = \mu_2 = \cdots = \mu_I$$
$$H_a : \text{not all of the } \mu_i \text{ are equal}$$

## Sums of Squares

Under $H_0$, we estimate the common value $\mu$ with the overall mean

$$\bar{x} = \sum_{i=1}^{I} \sum_{j=1}^{n_i} x_{ij}$$

What is left over $x_{ij} - \bar{x}$ is called the residual under $H_0$, and then the total residual variation in the data is given by

$$\text{SST} = \sum_{i=1}^{I} \sum_{j=1}^{n_i} (x_{ij} - \bar{x})^2$$

called the **total sum of squares.**

If we do not assume $H_0$ is true, then we should estimate each individual $\mu_i$ by the corresponding sample mean $\bar{x}_i$. The remainder $x_{ij} - \bar{x}_i$ is then the residual under an unrestricted model, and the total residual variation in the data is given by

$$\text{SSE} = \sum_{i=1}^{I} \sum_{j=1}^{n_i} (x_{ij} - \bar{x}_i)^2$$

called the **error**, or within groups, **sum of squares.** Intuitively, a better fit always occurs with the unrestricted model. We can show with a little algebra that the difference $\text{SSG} = \text{SST} - \text{SSE}$ is positive and can also be expressed as

$$\text{SSG} = \sum_{i=1}^{I} \sum_{j=1}^{n_i} (\bar{x}_i - \bar{x})^2$$

called the **between groups sum of squares.** Remarkably, it turns out that

$$\text{SST} = \text{SSG} + \text{SSE}$$

which is the key to the partition of the variation.

## The $F$ test

The magnitude of SSG measures the improvement in the fit as measured by the residual sum of squares. In order to reject $H_0$, the improvement must be significantly beyond what might be expected due to chance, and one is led to consider the ratio $\frac{\text{SSG}}{\text{SSE}}$. Define the mean squares

$$\text{MSG} = \frac{\text{SSG}}{I-1} \qquad \text{MSE} = \frac{\text{SSE}}{N-I}$$

where $N = \sum_{i=1}^{I} n_i$, and form the ratio

$$F = \frac{\text{MSG}}{\text{MSE}}$$

known to have an $F$ distribution with $I-1$ degrees of freedom for the numerator and $N-I$ degrees of freedom for the denominator (denoted by $F(I-1, N-I)$).

The decision rule is

$$\text{Reject } H_0 \text{ at level } \alpha \text{ if } F > F^*$$

where $F^*$ is the upper $\alpha$ critical value of an $F(I-1, N-I)$ distribution, that is, $F^*$ satisfies

$$P\left(F(I-1, N-I) > F^*\right) = \alpha$$

## Pooled Estimate of $\sigma^2$

We observe that MSE can also be expressed as

$$\text{SSE} = \sum_{i=1}^{I}(n_i - 1)s_i^2$$

where $\{s_i\}$ are the sample variances and, consequently, MSE is a pooled sample variance

$$s_p^2 \equiv \text{MSE} = \frac{\sum_{i=1}^{I}(n_i - 1)s_i^2}{\sum_{i=1}^{I}(n_i - 1)}$$

and therefore an unbiased estimate of $\sigma^2$.

## Coefficient of Determination

Finally, we define the ANOVA coefficient of determination

$$R^2 = \frac{\text{SSG}}{\text{SST}}$$

as the fraction of the total variance "explained by model $H_0$."

# Testing Hypotheses in a One-Way ANOVA

We illustrate the implementation of the preceding discussion with the following worked exercise.

> **Example 14.1.** (Exercise 14.61, page 14.50 in Companion Chapter 14.) The presence of harmful insects in farm fields is detected by erecting boards covered with a sticky material and then examining the insects trapped on each board. To investigate which colors are most attractive to cereal leaf beetles, researchers placed six boards of each of four colors in a field of oats. Table 14.1 gives data on the number of cereal leaf beetles trapped in July.
>
> (a) Make a table of means and standard deviations for the four colors, and plot the data and the means.
> (b) State $H_0$ and $H_a$ for an ANOVA on these data, and explain in words what ANOVA tests in this setting.
> (c) Using Excel, run the ANOVA. What are the $F$ statistic and its $P$-value? State the values of $s_p$ and $R^2$. What do you conclude?

Table 14.1: Luminescent Colors and Insect Attractiveness

| Color | Insects trapped | | | | | |
|-------|----|----|----|----|----|----|
| Lemon yellow | 45 | 59 | 48 | 46 | 38 | 47 |
| White | 21 | 12 | 14 | 17 | 13 | 17 |
| Green | 37 | 32 | 15 | 25 | 39 | 41 |
| Blue | 16 | 11 | 20 | 21 | 14 | 7 |

## Plotting the Data and the Sample Means

The first step in ANOVA is usually exploratory, where the data and means are plotted. Such a plot will help to confirm visually or dispel the assumption of equal variances and to indicate possible outliers or skewness in the data that might call into question use of this technique.

The step is easily carried out in Excel using the **ChartWizard**, which produces side-by-side displays of the samples $\{x_{ij}\}$, their sample means $\{\bar{x}_i\}$, and the overall mean $\bar{x}$, a good preliminary display of ANOVA data.

## Means and Standard Deviations

The mean and standard deviations are evaluated with the Excel function **AVERAGE** and **STDEV**. Refer to Figure 14.1 throughout.

1. Enter the data and labels in B3:E9 of a workbook.

2. Enter the label "mean" in A10 and then the formula = AVERAGE(B4:B9) in B10. The value 47.17 appears, which is the sample mean of the "Yellow" observations. Select cell B10 and, using the fill handle, drag to E10, filling the cells with the means for the other colors.

3. Enter the label "stdev" in A11 and the formula = STDEV(B4:B9) in B11. Then select B11 and drag the fill handle to E11. Now the standard deviations of all the samples appear.

### Plotting

The workbook needs to be prepared for the **ChartWizard** by relocating the data, coding the samples, relocating the sample means, and entering the overall mean. Complete columns G, H, I, J as indicated.

| | A | B | C | D | E | F | G | H | I | J |
|---|---|---|---|---|---|---|---|---|---|---|
| 1 | | | | One-Way Analysis of Variance | | | | | | |
| 2 | | | | | | | | | | |
| 3 | | Yellow | White | Green | Blue | | Color | Insects | Means | Overall |
| 4 | | 45 | 21 | 37 | 16 | | 1 | 45 | | |
| 5 | | 59 | 12 | 32 | 11 | | 1 | 59 | | |
| 6 | | 48 | 14 | 15 | 20 | | 1 | 48 | | |
| 7 | | 46 | 17 | 25 | 21 | | 1 | 46 | | |
| 8 | | 38 | 13 | 38 | 14 | | 1 | 38 | | |
| 9 | | 47 | 17 | 41 | 7 | | 1 | 47 | | |
| 10 | mean | 47.17 | 15.67 | 31.33 | 14.83 | | 2 | 21 | | |
| 11 | stdev | 6.795 | 3.327 | 9.771 | 5.345 | | 2 | 12 | | |
| 12 | | | | | | | 2 | 14 | | |
| 13 | | | | | | | 2 | 17 | | |
| 14 | | | | | | | 2 | 13 | | |
| 15 | | | | | | | 2 | 17 | | |
| 16 | | | | | | | 3 | 37 | | |
| 17 | | | | | | | 3 | 32 | | |
| 18 | | | | | | | 3 | 15 | | |
| 19 | | | | | | | 3 | 25 | | |
| 20 | | | | | | | 3 | 38 | | |
| 21 | | | | | | | 3 | 41 | | |
| 22 | | | | | | | 4 | 16 | | |
| 23 | | | | | | | 4 | 11 | | |
| 24 | | | | | | | 4 | 20 | | |
| 25 | | | | | | | 4 | 21 | | |
| 26 | | | | | | | 4 | 14 | | |
| 27 | | | | | | | 4 | 7 | | |
| 28 | | | | | | | 1 | | 47.17 | 27.25 |
| 29 | | | | | | | 2 | | 15.67 | 27.25 |
| 30 | | | | | | | 3 | | 31.33 | 27.25 |
| 31 | | | | | | | 4 | | 14.83 | 27.25 |

Figure 14.1: Preparing the Data for the One-Way ANOVA Tool

Because the sample means have already been calculated in B10:E10, an efficient way to relocate them to I28:I31 is to select B10:E10, choose **Edit − Copy** from the Menu Bar, then select cell I28 and choose **Edit − Paste Special** from the Menu Bar. In the **Paste Special** dialog box, check the radio button for **values** because the contents of B10:E10 are formulas, not values, and a straight **Edit − Copy** will change the relative cell references. Then check the box **Transpose** to convert this row selection into a column and click **OK**. Finally, the common value 27.25 in J28:J31 is the overall mean obtained from the Excel formula = AVERAGE(B4:E9).

**Users of Excel 5/95**

Begin as always by clicking the **ChartWizard** button.

- In Step 1 of 5 enter G3:J31 for the range. This will produce a simultaneous scatterplot with all three variables—data, sample means, overall mean— plotted on the $y$-axis (with different markers) against the colors (column G) on the $x$-axis.

- In Step 2 select **XY (Scatter)**.

- In Step 3 select Format **1**.

- In Step 4 use **Columns** for Data Series in, First "1" Column for X Data, and First "1" Row for Legend Text.

- In Step 5 select **Yes** for **Add a Legend?**, type "Plot of the Data and the Means" for the **Chart Title**, type "Color" for **Axis Title Category (X)**, and type "Insects Trapped" for **Axis Title Value (Y)**. Click **Finish**.

**Users of Excel 97/98/2000/2001**

- In Step 1 select **XY (Scatter)** for **Chart Type** and the upper left Chart sub-type **Scatter**.

- In Step 2 under the **Data Range** tab enter G3:J31 (which will produce a simultaneous scatterplot with all three variables – data, sample means, and overall mean – plotted on the $y$-axis (with different markers) against the colors (column G) on the $x$-axis, and select the **Series** radio button for **Columns**.

- In Step 3 under the **Titles** tab, type "Plot of the Data and the Means" as the **Chart Title**, "Color" as **Value (X) Axis**, and "Insects Trapped" as **Value (Y) Axis**. Under the **Axes** tab, both check boxes should be selected. Under the **Gridlines** tab, clear all check boxes. Under the **Legend** tab, check the **Show legend** and locate it to the right. Finally, under the **Data Labels** tab, select the radio button **None**.

- In Step 4 embed the graph in the current workbook by selecting the radio button **As object in**. Click **Finish**.

A scatterplot like the one shown in Figure 14.2 appears with distinct markers representing the individual observations, the sample means, and the overall mean. Editing will enhance the usefulness of this plot.

**Editing the Plot**

We edit Figure 14.2 to emphasize more clearly the observations, the sample means, and the overall mean. The result resembles side-by-side boxplots.

1. Activate the chart for editing.

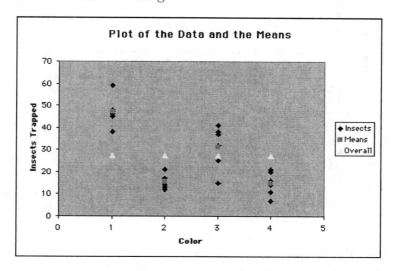

Figure 14.2: Default Plot—Data and Means

2. Click one of the markers representing a sample mean (the entire series will then appear highlighted). From the Menu Bar, choose **Format – Selected Data Series....** In the dialog box under the **Patterns** tab, select radio buttons **Automatic** for **Line** and **None** for **Marker**. Click **OK**. The markers for sample means are now replaced by a connecting polygonal line.

3. Repeat this step after selecting a marker for the overall mean. The four markers are replaced by a horizontal line.

4. Click a marker for the observations and change the **Style** of the marker to a square.

5. Select the horizontal axis, and from the Menu Bar choose **Format – Selected Axis....** In the dialog box, click the **Scale** tab. Then type "1" for **Minimum**, "4" for **Maximum**, "1" for **Major unit**, and "1" for **Minor unit**. Then click the **Patterns** tab, click **None** for **Tick Mark Labels**, and click **None** for **Major** and **Minor** Tick Mark Type. Click **OK**.

6. Activate the **Plot Area** and change the background color to white using the color swatch. Click OK.

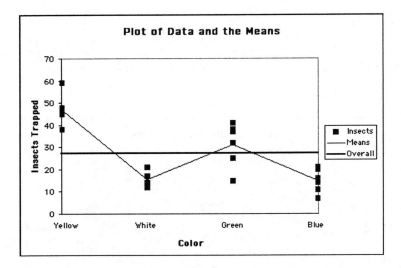

Figure 14.3: Enhanced Plot—Data and Means

7. Type the word "Yellow" in the **Formula Bar** and press enter. The word "Yellow" appears in a grey shaded text rectangle. With the cursor, drag it to the location shown in Figure 14.3. Click outside the text rectangle to deselect it. Now repeat for the other three labels "White," "Green," and "Blue."

The result of this enhancement is Figure 14.3, which immediately shows the dominant features of the data set much more aggressively than the numerical counterpart in cells B3:E11 of Figure 14.1.

## Using the One-Way ANOVA Tool

1. From the Menu Bar choose **Tools – Data Analysis** and select **Anova: Single Factor** from the tools listed in the **Data Analysis** dialog box. Click **OK** to display the **Anova: Single Factor** dialog box (Figure 14.4).

2. Type the cell references B3:E9 (or point and drag on the workbook) for **Input range**. Select the radio button **Grouped By: Columns**. Check the box **Labels in first row** and use "0.05" for **Alpha**. Enter K1 for **Output range**. Click **OK**.

## Excel Output—ANOVA Table

The output from the **Anova: Single Factor** tool appears in Figure 14.5, from which we can read off all the variables described earlier in this chapter. The first

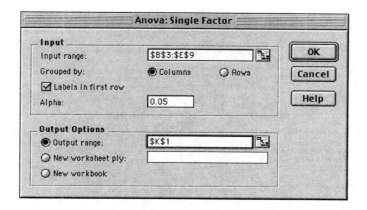

Figure 14.4: Single-Factor ANOVA Dialog Box

| | K | L | M | N | O | P | Q |
|---|---|---|---|---|---|---|---|
| 3 | Anova: Single Factor | | | | | | |
| 4 | | | | | | | |
| 5 | SUMMARY | | | | | | |
| 6 | *Groups* | *Count* | *Sum* | *Average* | *Variance* | | |
| 7 | Yellow | 6 | 283 | 47.167 | 46.167 | | |
| 8 | White | 6 | 94 | 15.667 | 11.067 | | |
| 9 | Green | 6 | 188 | 31.333 | 95.467 | | |
| 10 | Blue | 6 | 89 | 14.833 | 28.567 | | |
| 11 | | | | | | | |
| 12 | | | | | | | |
| 13 | ANOVA | | | | | | |
| 14 | *Source of Variation* | *SS* | *df* | *MS* | *F* | *P-value* | *F crit* |
| 15 | Between Groups | 4210.167 | 3 | 1403.389 | 30.97 | 1.03E-07 | 3.10E+00 |
| 16 | Within Groups | 906.333 | 20 | 45.317 | | | |
| 17 | | | | | | | |
| 18 | Total | 5116.5 | 23 | | | | |

Figure 14.5: Single-Factor ANOVA Output

half of the output provides summary statistics for the four samples. Had we not desired a plot we would have read the sample means and standard deviations directly from this output.

### Source of Variation

Between Groups sum of squares refers to

$$\text{SSG} = 4210.167$$

Within Groups (error) sum of squares is

$$\text{SSE} = 906.333$$

The Total sum of squares is

$$\text{SST} = 5116.5$$

Further, we read the degrees of freedom ($df$)

$$\text{DFG} = 3 \quad \text{DFE} = 20 \quad \text{DFT} = 23$$

Consequently, the mean squares ($MS$) are

$$\text{MSG} = \frac{4210.167}{3} = 1403.389$$
$$\text{MSE} = \frac{906.333}{20} = 45.317$$

and the calculated $F$ statistic is

$$F = \frac{\text{MSG}}{\text{MSE}} = \frac{1403.389}{45.317} = 30.97$$

The 5% critical $F$ value ($F$ crit) is

$$F^* = 3.098$$

The $P$-value is 0.00000.

We conclude that for the significance test

$$H_0 : \mu_1 = \mu_2 = \mu_3 = \mu_4$$
$$H_a : \text{ not all of the } \mu_i \text{ are equal}$$

We reject $H_0$ at the 5% level (in fact at *any* reasonable level) in view of the small $P$-value.

We observe that the pooled estimate of variance is

$$s_p^2 = \text{MSE} = 45.17$$

which we note is also (because of the equal sample sizes) the average of the individual sample variances.

Finally, although the coefficient of determination is not provided, we can do the arithmetic to derive

$$R^2 = \frac{\text{SSG}}{\text{SST}} = \frac{4210.167}{5116.5} = 0.823$$

# Chapter 15

# Two-Way Analysis of Variance

In a one-way ANOVA, independent samples are taken from $I$ populations each differing with respect to one categorical variable, the population mean. We can view this variable as representing the levels of a particular factor as discussed in Chapter 3 of the text. In this setting, the one-way ANOVA then describes the analysis of a **completely randomized design**.

## 15.1 The Two-Way ANOVA Model

Another design discussed in Chapter 3 was a **randomized block design**. We recall that a block is a group of experimental units similar in some way that is expected to influence the response. The paired two-sample $t$ procedure is an example of a block design analysis.

Two-way ANOVA may then be viewed as a generalization of the paired $t$ test when there are more than two populations. The approach parallels one-way ANOVA by partitioning the total variation into components that can be interpreted as representing the contributions of factor effects or block effects.

However, the analysis of a randomized block design also applies when there are two or more factors whose effect is of interest, where one is not necessarily a blocking variable and where there are multiple observations (replications) for each combination of factor levels. That is the model we consider here.

The data comprise samples of size $n_{ij}$ for $I \times J$ treatments representing the combinations of I levels of factor $A$ and $J$ levels of factor $B$. The data are represented by $\{x_{ijk} : 1 \leq i \leq I,\ 1 \leq j \leq J,\ 1 \leq k \leq n_{ij}\}$, where $x_{ijk}$ represents the $k$th observation of treatment combination $(i, j)$. The model states

$$x_{ijk} = \mu_{ij} + \varepsilon_{ijk} \quad 1 \leq I,\ 1 \leq j \leq J,\ 1 \leq k \leq n_{ij}$$

where $\{\varepsilon_{ijk}\}$ are independent $N(0, \sigma)$ random variables. Note the assumption of common variance, as with the one-way model.

First, we view the model as a one-way ANOVA for $I \times J$ populations (treatments).

$$\texttt{DATA} = \texttt{FIT} + \texttt{RESIDUAL}$$

In particular, we can estimate an overall mean $\mu$ by

$$\bar{x} = \frac{1}{N} \sum_{i=1}^{I} \sum_{j=1}^{J} \sum_{k=1}^{n_{ij}} x_{ijk}$$

and then produce a corresponding **total sum of squares** where $N$ is the overall total number of observations.

$$\text{SST} = \sum_{i=1}^{I} \sum_{j=1}^{J} \sum_{k=1}^{n_{ij}} (x_{ijk} - \bar{x})^2$$

Likewise, we can estimate the residual or error variation, using the sample means for each $(i, j)$ combination of $I \times J$ treatments

$$\bar{x}_{ij} = \frac{1}{n_{ij}} \sum_{k=1}^{n_{ij}} x_{ijk}$$

and then we have the **error (residual) sum of squares**

$$\text{SSE} = \sum_{i=1}^{I} \sum_{j=1}^{J} \sum_{k=1}^{n_{ij}} (x_{ijk} - \bar{x}_{ij})^2$$

As with the one-way ANOVA this formula can be re-expressed as

$$\text{SSE} = \sum_{i=1}^{I} \sum_{j=1}^{J} (n_{ij} - 1) s_{ij}^2$$

where $s_{ij}^2$ is the sample variance for the $(i, j)$ combination, which leads to a pooled estimate for the common variance $\sigma^2$

$$s_p^2 = \frac{\text{SSE}}{\sum_{i=1}^{I} \sum_{j=1}^{J} (n_{ij} - 1)} = \frac{\text{SSE}}{N - IJ}$$

The corresponding **between groups sum of squares** is here denoted (with a change in notation, SSM replacing SSG) as

$$\text{SSM} = \sum_{i=1}^{I} \sum_{j=1}^{J} \sum_{k=1}^{n_{ij}} (\bar{x}_{ij} - \bar{x})^2$$

To this stage, the analysis is for a one-way ANOVA. But because of the way the data were collected (the design), it is then possible to partition SSM further into additional sums of squares that can be identified as arising from model parameters. We therefore prescribe $\mu_{ij}$ using a **linear model**

$$\mu_{ij} = \mu + \alpha_i + \beta_j + \gamma_{ij}$$

where $\mu$ is an overall mean, $\alpha_i$ represents an effect due to level $i$ of factor $A$, $\beta_j$ represents an effect due to level $j$ of factor $B$ (both of these are called **main effects**), and $\gamma_{ij}$ represents an **interaction** effect between factor $A$ and factor $B$. We say that two factors interact if the difference in mean response for two levels of one factor is not constant across levels of the second factor.

Excel provides two versions of **two-way ANOVA** in the **Analysis ToolPak**. The first is where $n_{ij} = 1$ for all $(i, j)$ (in which case $\gamma_{ij}$ cannot be estimated). The other is where $n_{ij} \equiv n \geq 2$, so there is replication of observations at each $(i, j)$ level but the number of replications is the same. This is called a **balanced** design. With unequal sample sizes, the ANOVA formulas become more complex and the factor effect components (sums of squares) are no longer orthogonal (they don't add up). For this reason, we now limit the discussion to balanced designs with the same number of observations $n_{ij} \equiv n$ per treatment.

The parameters in the representation of $\mu_{ij}$

$$\mu_{ij} = \mu + \alpha_i + \beta_j + \gamma_{ij}$$

are not uniquely determined, because we can add and subtract constants on the right-hand side, changing the parameters but maintaining equality. It is necessary to impose constraints for uniqueness, namely,

$$\sum_{i=1}^{I} \alpha_i = 0, \quad \sum_{j=1}^{J} \beta_j = 0, \quad \sum_{i=1}^{I} \gamma_{ij} = 0, \quad \sum_{j=1}^{I} \gamma_{ij} = 0$$

We can then **interpret** $\alpha_i$ as the contribution or deviation of level $i$ of factor $A$ from a baseline (of 0 in view of $\sum_{i=1}^{I} \alpha_i = 0$). Likewise, $\beta_j$ is the contribution of level $j$ of factor $B$, and $\gamma_{ij}$ is any possible interaction of combination $(i, j)$.

If we fix attention on level $i$ of factor $A$ and average over all levels of factor $B$, we can estimate $\mu + \alpha_i$ using the sample mean of all observations with the same value $i$,

$$\frac{1}{Jn} \sum_{j=1}^{J} \sum_{k=1}^{n} x_{ijk} = \bar{x}_{i\bullet}$$

We can also estimate $\mu + \beta_j$ using

$$\frac{1}{In} \sum_{i=1}^{I} \sum_{k=1}^{n} x_{ijk} = \bar{x}_{\bullet j}$$

We then take differences and find the natural estimates (recalling that $\bar{x}$ is the overall mean):

$$\hat{\alpha}_i \equiv \bar{x}_{i\bullet} - \bar{x} \qquad \text{estimates } \alpha_i$$

$$\hat{\beta}_i \equiv \bar{x}_{\bullet j} - \bar{x} \qquad \text{estimates } \beta_j$$

$$\hat{\gamma}_{ij} \equiv \bar{x}_{ij} - \bar{x}_{i\bullet} - \bar{x}_{\bullet j} + \bar{x} \quad \text{estimates } \gamma_{ij}$$

Then a minor miracle occurs. Each of these estimates contributes to a marvelous sum of squares decomposition; namely, if we define

$$\text{SSA} = \sum_{i=1}^{I} \sum_{j=1}^{J} \sum_{k=1}^{n} \hat{\alpha}_i^2$$

$$\text{SSB} = \sum_{i=1}^{I} \sum_{j=1}^{J} \sum_{k=1}^{n} \hat{\beta}_j^2$$

$$\text{SSAB} = \sum_{i=1}^{I} \sum_{j=1}^{J} \sum_{k=1}^{n} \hat{\gamma}_{ij}^2$$

then

$$\text{SSM} = \text{SSA} + \text{SSB} + \text{SSAB}$$

Each term on the right carries its own degrees of freedom

$$\text{DFA} = I - 1$$

$$\text{DFB} = J - 1$$

$$\text{DFAB} = (I-1)(J-1)$$

giving corresponding mean squares and $F$ ratios as in one-way ANOVA. For instance,

$$\text{MSA} = \frac{\text{SSA}}{I-1}$$

$$F = \frac{\text{MSA}}{\text{MSE}}$$

and this particular $F$ ratio is used to test

$$H_{0A} : \alpha_i = 0, \quad 1 \leq i \leq I$$

$$H_{aA} : \text{ at least two } \alpha_i \text{ are not zero}$$

The decision rule is that if

$$F = \frac{\text{MSA}}{\text{MSE}} \quad \text{exceeds the critical value } F^*(I-1, N - IJ)$$

then we reject $H_{0A}$ and conclude that there is a difference among the means for the levels of factor $A$; that is, factor $A$ is significant. An analogous analysis is carried out for factor $B$ and then for interaction if neither factor $A$ nor factor $B$ is judged significant.

## 15.2 Inference for Two-Way ANOVA

While the theory just described may seem intimidating, the practical implementation of the procedure could not be simpler, once the data are properly recorded in the workbook. Excel outputs summary statistics, sums of squares, mean squares, $F$ ratios, critical $F$ values, and $P$-values in an extensive ANOVA table.

**Example 15.1.** (Exercise 15.30, page 15.29 in Companion Chapter 15.) Iron deficiency anemia is the most common form of malnutrition in developing countries, affecting about 50% of children and women, and 25% of men. Iron pots for cooking foods had traditionally been used in many of these countries, but they have been largely replaced by aluminum pots, which are cheaper and lighter. Some research has suggested that food cooked in iron pots will contain more iron than food cooked in other types of pots. One study designed to investigate this issue compared the iron content of some Ethiopian foods cooked in aluminum, clay, and iron pots. The iron content was measured of *yesiga wet'*, beef cut into small pieces and prepared with several Ethiopian spices; *shiro wet'*, a legume-based mixture of chickpea flour and Ethiopian spiced pepper; and *ye-atkilt allych'a*, a lightly spiced vegetable casserole. In the table below, these three foods are labeled meat, legumes, and vegetables. Four samples of each food were cooked in each type of pot. The iron in the food is measured in milligrams of iron per 100 grams of cooked food. Here are the data.

| | Iron Content | | | | | | | | | | | |
|---|---|---|---|---|---|---|---|---|---|---|---|---|
| Type of Pot | Meat | | | | Legumes | | | | Vegetables | | | |
| Aluminum | 1.77 | 2.36 | 1.96 | 2.14 | 2.40 | 2.17 | 2.41 | 2.34 | 1.03 | 1.53 | 1.07 | 1.30 |
| Clay | 2.27 | 1.28 | 2.48 | 2.68 | 2.41 | 2.43 | 2.57 | 2.48 | 1.55 | 0.79 | 1.68 | 1.82 |
| Iron | 5.27 | 5.17 | 4.06 | 4.22 | 3.69 | 3.43 | 3.84 | 3.72 | 2.45 | 2.99 | 2.80 | 2.92 |

(a) Make a table giving the sample size, mean, and standard deviation for each type of pot.
(b) Plot the means and give a short summary of how the iron content of foods depends upon the cooking pot.
(c) Run the analysis of variance. Give the ANOVA table, the $F$ statistics with degrees of freedom and $P$-values, and your conclusions regarding the hypotheses about main effects and interactions.

## The Two-Way ANOVA Tool

The explanatory variable of interest is the type of pot, while another variable, type of food, has been used as a blocking variable. This is equivalent to a two-way anova and we will therefore refer to "Pot" as Factor $A$ (3 levels: Aluminum, Clay, Iron) and "Food" as Factor $B$ (three levels: Meat, Legumes, Vegetables). There are four replications per treatment, so $n = 4$ for a total of $N = I \times J \times n = 3 \times 3 \times 4 = 36$ observations.

| | A | B | C | D |
|---|---|---|---|---|
| 1 | **Two-Way Analysis of Variance** | | | |
| 2 | | | | |
| 3 | Pot | | Food | |
| 4 | | Meat | Legumes | Vegetables |
| 5 | Aluminum | 1.77 | 2.40 | 1.03 |
| 6 | | 2.36 | 2.17 | 1.53 |
| 7 | | 1.96 | 2.41 | 1.07 |
| 8 | | 2.14 | 2.34 | 1.30 |
| 9 | Clay | 2.27 | 2.41 | 1.55 |
| 10 | | 1.28 | 2.43 | 0.79 |
| 11 | | 2.48 | 2.57 | 1.68 |
| 12 | | 2.68 | 2.48 | 1.82 |
| 13 | Iron | 5.27 | 3.69 | 2.45 |
| 14 | | 5.17 | 3.43 | 2.99 |
| 15 | | 4.06 | 3.84 | 2.80 |
| 16 | | 4.22 | 3.72 | 2.92 |

Figure 15.1: Preparing the Data

## Preparing the Data

1. Enter the data exactly as shown in Fig 15.1. Excel will balk if the data are not laid out correctly.

2. From the Menu Bar, choose **Tools – Data Analysis – Anova: Two-Factor With Replication** and complete the dialog box, with entries as shown in Fig 15.2. The output appears in cells E1:K36.

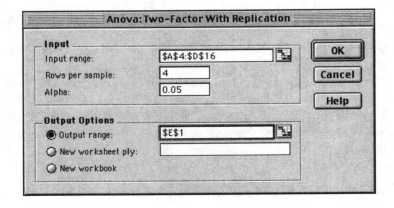

Figure 15.2: Two-Way ANOVA Dialog Box

| | E | F | G | H | I | J | K |
|---|---|---|---|---|---|---|---|
| 1 | Anova: Two-Factor With Replication | | | | | | |
| 2 | | | | | | | |
| 3 | SUMMARY | Meat | Legumes | Vegetables | Total | | |
| 4 | *Aluminum* | | | | | | |
| 5 | Count | 4 | 4 | 4 | 12 | | |
| 6 | Sum | 8.23 | 9.32 | 4.93 | 22.48 | | |
| 7 | Average | 2.0575 | 2.3300 | 1.2325 | 1.8733 | | |
| 8 | Variance | 0.063492 | 0.012333 | 0.053492 | 0.272770 | | |
| 9 | | | | | | | |
| 10 | *Clay* | | | | | | |
| 11 | Count | 4 | 4 | 4 | 12 | | |
| 12 | Sum | 8.71 | 9.89 | 5.84 | 24.44 | | |
| 13 | Average | 2.1775 | 2.4725 | 1.4600 | 2.0367 | | |
| 14 | Variance | 0.386025 | 0.005092 | 0.211667 | 0.361606 | | |
| 15 | | | | | | | |
| 16 | *Iron* | | | | | | |
| 17 | Count | 4 | 4 | 4 | 12 | | |
| 18 | Sum | 18.72 | 14.68 | 11.16 | 44.56 | | |
| 19 | Average | 4.6800 | 3.6700 | 2.7900 | 3.7133 | | |
| 20 | Variance | 0.394733 | 0.029800 | 0.057533 | 0.781970 | | |
| 21 | | | | | | | |
| 22 | *Total* | | | | | | |
| 23 | Count | 12 | 12 | 12 | | | |
| 24 | Sum | 35.66 | 33.89 | 21.93 | | | |
| 25 | Average | 2.9717 | 2.8242 | 1.8275 | | | |
| 26 | Variance | 1.824724 | 0.406808 | 0.602730 | | | |
| 27 | | | | | | | |
| 28 | | | | | | | |
| 29 | ANOVA | | | | | | |
| 30 | *Source of Variation* | SS | df | MS | F | P-value | F crit |
| 31 | Sample | 24.89396 | 2 | 12.44698 | 92.263 | 0.00000 | 3.354 |
| 32 | Columns | 9.29687 | 2 | 4.64844 | 34.456 | 0.00000 | 3.354 |
| 33 | Interaction | 2.64043 | 4 | 0.66011 | 4.893 | 0.00425 | 2.728 |
| 34 | Within | 3.6425 | 27 | 0.13491 | | | |
| 35 | | | | | | | |
| 36 | Total | 40.47376 | 35 | | | | |

Figure 15.3: Two-Way ANOVA Output

## Excel Output—Two-Way ANOVA

There are two components to the output in Fig 15.3. The top portion provides summary statistics such as sample means and variances, which are useful in plotting. The lower half is the ANOVA table.

## Summary

Referring to Fig 15.3, the block E3:H20 gives summary statistics for each treatment labeled by a level of Factor A in column E and a level of Factor B in row 3. For instance, for the treatment (Aluminum × Meat) we read in cell F5 there are four observations, from cell F7 the average $x_{11} = 2.0575$, and from cell F8 the variance $s_{11}^2 = 0.063492$, and so on. Standard deviations are not given but may be calculated from the variances.

The column block on the right labeled "Total" (I3:G20) provides summaries for each level of factor $A$ (summed or averaged over all levels of factor $B$). For instance, the average iron content for Iron pots is shown in cell I19 as $\bar{x}_{3\bullet} = 3.7133$. This is the average of the 12 observations corresponding to iron pots. Likewise, the average iron content for Aluminum pots is $\bar{x}_{1\bullet} = 1.8733$ in cell I7, and for Clay pots the average iron content is $\bar{x}_{2\bullet} = 2.0367$ in cell I13.

The corresponding summaries for factor $B$ are in the row block headed by the label Total in E22:E26. For example, the average for the 12 observations for "Meat" is $\bar{x}_{\bullet 1} = 2.9717$ in cell F25.

## ANOVA Table

There are three separate possible significance tests: Factor $A$, Factor $B$, and Interaction; Block E29:K36 presents the ANOVA table listing the four sources of variation (with the terminology: Sample $\equiv$ Factor $A$; Column $\equiv$ Factor $B$; Interaction, Within $\equiv$ Error; and Total), their corresponding sums of squares, degrees of freedom, mean square, computed $F$, $P$-value, and critical $F^*$ values.

We can immediately read off the conclusions. First, there does not appear to be any significant interaction effect, so we may then examine the two factors individually. For Factor $A$, because the corresponding computed $F$ statistic 92.263 exceeds the critical $F^*(2, 27) = 3.354$, we conclude that there is a difference in iron content among the three types of pots. We also conclude that there is a difference in iron content due to Factor $B$, the type of food. This was to be expected and shows that blocking on this factor was effective.

## Profile (Interaction) Plot

A profile plot is a simple graphical diagnostic tool for displaying the numerical summaries in the Excel output. It is handy for seeing possible interactions visually. A profile plot is a graph of all the treatment means in a manner that gives some visual insight. The sample means of all the treatments are plotted on the $y$-axis against the corresponding levels of one of the factors on the $x$-axis. The following steps show how to use the **ChartWizard** and the summary output to produce a profile plot. While it really does not matter which factor is used for $x$, because the same information is displayed in either case, yet some features may be more apparent visually in one graph than in the other. Because "Pot" is the primary factor of interest we will plot the sample means against the types of pots on the $x$ axis.

1. Enter the data as in the top half of Fig 15.4.

2. **Users of Excel 5/95**

   - Click the **ChartWizard** button. In Step 1 enter the range N4:Q7, in Step 2 select **Line** chart, in Step 3 select **Format 1**, in Step 4 select **Data Series in Rows, Row 1** for **Category(X) Labels, Use**

| | N | O | P | Q |
|---|---|---|---|---|
| 1 | | Profile Plot of Means | | |
| 2 | | | | |
| 3 | Pot | | Food | |
| 4 | | Meat | Legumes | Vegetables |
| 5 | Aluminum | 2.058 | 2.330 | 1.233 |
| 6 | Clay | 2.178 | 2.473 | 1.460 |
| 7 | Iron | 4.680 | 3.670 | 2.790 |

Figure 15.4: Interactions—Data

Column 1 for **Legend Text**, and finally in Step 5 select **Yes** for **Add a Legend?** and type the name for the chart and the axis titles.

### Users of Excel 97/98/2000/2001

- Select the range N4:Q7 and click the **ChartWizard** button. In Step 1 select **Line** for **Chart Type** and the upper left Chart sub-type **Line**. In Step 2 under the **Data Range** tab, enter K3:O6 and select the **Series** radio button for **Rows**: (Note what happens in the sample display if you select the radio button **Columns**.) In Step 3 enter the titles under the **Titles** tab, remove gridlines under the **Gridlines** tab, check **Show legend** under the **Legend** tab, and under the **Data Labels** tab select the radio button **None**. In Step 4 click **Finish**.

3. Finally, make any additional editing changes so that the resulting chart appears similar to Fig 15.5.

### Output

The profile plot shows that the sample means for iron pots are all larger than the sample means for the other types of pots. Also notice that there is a crossover in the lines. Legumes, which show a consistently higher iron content than meat or vegetables in aluminum and clay pots, are second highest in iron content in the iron pots. Such a pattern indicates the possibility that interaction is present. However, the $F$ test for ANOVA did not indicate this interaction. The apparent contradiction results because the estimated standard deviation ($s_p^2$ from cell H34) $s_p = \sqrt{0.13491} = 0.367$ is so large that such a crossover in means can be explained as random variation. This example indicates that some caution needs to be exercised in interpreting profile plots.

**Exercise.** By selecting the **Series** radio button for **Rows** in the **ChartWizard**, produce the profile plot shown in Fig 15.6, with type of food on the horizontal axis. The same nine sample means are plotted, but with different abscissae than in Fig 15.5.

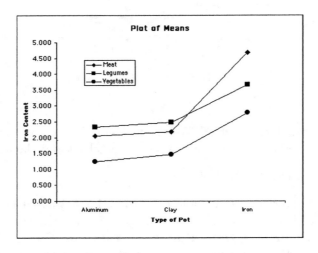

Figure 15.5: Profile Plot vs. Pots

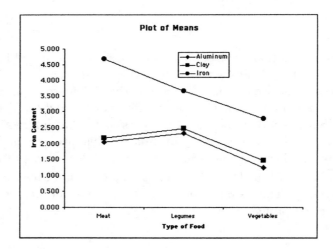

Figure 15.6: Profile Plot vs. Food

# Chapter 16

# Nonparametric Tests

Nonparametric procedures are based on the ranks of observations and replace assumptions of normality with less stringent distributional assumptions, such as symmetry and continuity. Excel does not provide nonparametric tests. Nonetheless, these tests may be readily developed.

## 16.1   The Wilcoxon Rank Sum Test

The Wilcoxon rank sum test is a procedure for comparing independent samples from two populations. Suppose

$$(x_{11}, x_{12}, \ldots, x_{1n_1}), (x_{21}, x_{22}, \ldots, x_{2n_2})$$

are the two samples. Generally, sample 1, the $x_{1j}$ observations, might be the control group, while sample 2, the $x_{2j}$ observations, might be the treatment group.

We wish to test whether both samples can be assumed to arise from identical populations or whether the populations are shifted by a constant. The precise assumption is that if $F_i$ represents the cumulative distribution function of the $x_i$-observations, $1 \leq i \leq 2$, then

$$F_2(t) = F_1(t + \Delta) \qquad \text{for all real } t$$

where $\Delta$ is a constant representing an unknown shift in the two distributions. If $\Delta$ is positive, then the second sample would contain systematically higher values than the first sample.

The null hypothesis is thus

$$H_0 : \mu_1 - \mu_2 = 0$$

where $\mu_1$ and $\mu_2$ are the two population means (or medians).

It is also possible to test a nonzero difference

$$H_0 : \mu_1 - \mu_2 = \Delta_0$$

by subtracting $\Delta_0$ from each sample 1 observation and applying the first test for a zero difference in means to

$$(x_{11} - \Delta_0, x_{12} - \Delta_0, \ldots, x_{1n_1} - \Delta_0), (x_{21}, x_{22}, \ldots, x_{2n_2})$$

## Description of the Procedure

1. Rank all the observations from smallest to largest in one list with $N = n_1 + n_2$ observations.

2. Let $r_j$ be the rank of the observation $x_{1j}$.

3. Set

$$W = \sum_{j=1}^{n_1} r_j$$

   which is the sum of the ranks associated with the data from sample 1.

If $H_0$ is true, then each overall ranking of the $N$ combined observations would have the same probability, but if, for instance,

$$H_a : \mu_1 - \mu_2 > 0$$

then the ranks of sample 1, contributing to $W$ and arising from the population with the larger mean under $H_a$, would be larger than expected under $H_0$, leading to an observed $W$ value above the mean. The Wilcoxon rank sum test therefore rejects $H_0$ if $W$ is beyond some reasonable value. This test is sometimes called the Mann-Whitney test because it was originally derived as an alternative, but equivalent, formulation by H. B. Mann and D. R. Whitney in 1947.

The exact distribution of $W$ has been tabulated, but we will base our procedure on the normal approximation. It can be shown that if $H_0$ is true, then

$$\mu_W = \text{ mean of } W = \frac{n_1(n_1 + n_2 + 1)}{2}$$

$$\sigma_w = \text{ standard deviation of } W = \sqrt{\frac{n_1 n_2 (n_1 + n_2 + 1)}{12}}$$

Calculate

$$z = \frac{W - \mu_W}{\sigma_W}$$

For a fixed level $\alpha$ test:

$$\text{if } H_a : \mu_1 - \mu_2 > 0 \quad \text{reject } H_0 \text{ if } z > z^*$$
$$\text{if } H_a : \mu_1 - \mu_2 < 0 \quad \text{reject } H_0 \text{ if } z < -z^*$$
$$\text{if } H_a : \mu_1 - \mu_2 \neq 0 \quad \text{reject } H_0 \text{ if } |z| > z^*$$

where $z^*$ represents the corresponding upper critical value of a standard normal distribution.

# The Wilcoxon Rank Sum Test in Practice

| | A | B | C | D | E | F |
|---|---|---|---|---|---|---|
| 1 | | | Wilcoxon Rank Sum Test | | | |
| 2 | Asia | Eastern Europe | Population | Combined | Rank | |
| 3 | 2.3 | 6.0 | 2 | -12.1 | 1 | |
| 4 | 8.8 | -5.2 | 2 | -5.2 | 2 | |
| 5 | 3.9 | -1.0 | 2 | -1.7 | 3 | |
| 6 | 4.1 | 3.5 | 2 | -1.5 | 4 | |
| 7 | 6.4 | 3.2 | 2 | -1.0 | 5 | |
| 8 | 5.9 | -1.7 | 2 | 1.0 | 6 | |
| 9 | 4.2 | -1.5 | 1 | 1.3 | 7 | |
| 10 | 2.9 | 1.0 | 2 | 1.4 | 8 | |
| 11 | 1.3 | 4.9 | 1 | 2.3 | 9 | |
| 12 | 5.1 | 1.4 | 1 | 2.9 | 10 | |
| 13 | 5.6 | 7.0 | 2 | 3.2 | 11 | |
| 14 | 6.2 | 3.7 | 2 | 3.5 | 12 | |
| 15 | | -12.1 | 2 | 3.7 | 13 | |
| 16 | | | 1 | 3.9 | 14 | |
| 17 | | | 1 | 4.1 | 15 | |
| 18 | | | 1 | 4.2 | 16 | |
| 19 | | | 2 | 4.9 | 17 | |
| 20 | | | 1 | 5.1 | 18 | |
| 21 | | | 1 | 5.6 | 19 | |
| 22 | | | 1 | 5.9 | 20 | |
| 23 | | | 2 | 6.0 | 21 | |
| 24 | | | 1 | 6.2 | 22 | |
| 25 | | | 1 | 6.4 | 23 | |
| 26 | | | 2 | 7.0 | 24 | |
| 27 | | | 1 | 8.8 | 25 | |
| 28 | | | | | | |
| 29 | alpha | 0.05 | | | | |
| 30 | Calculations | | | | | |
| 31 | n_1 | 12 | =COUNT(Asia) | | | |
| 32 | n_2 | 13 | =COUNT(Eastern Europe) | | | |
| 33 | W | 198 | =SUMIF(Population, "=1", Rank) | | | |
| 34 | mu | 156 | =n_1*(n_1+n_2+1)/2 | | | |
| 35 | sigma | 18.385 | =SQRT(n_1*n_2*(n_1+n_2+1)/12) | | | |
| 36 | z | 2.284 | =(W-mu)/sigma | | | |
| 37 | Lower Test | | | | | |
| 38 | lower_z | | =NORMSINV(alpha) | | | |
| 39 | Decision | | =IF(z<lower_z,"Reject H0","Do Not Reject H0") | | | |
| 40 | Pvalue | | =NORMSDIST(z) | | | |
| 41 | Upper Test | | | | | |
| 42 | upper_z | 1.645 | =-NORMSINV(alpha) | | | |
| 43 | Decision | Reject H0 | =IF(z>upper_z,"Reject H0","Do Not Reject H0") | | | |
| 44 | Pvalue | 0.0112 | =1-NORMSDIST(z) | | | |
| 45 | Two-Sided Test | | | | | |
| 46 | two_z | | =ABS(NORMSINV(alpha/2)) | | | |
| 47 | Decision | | =IF(ABS(z)>two_z,"Reject H0","Do Not Reject H0") | | | |
| 48 | Pvalue | | =2*(1-NORMSDIST(ABS(z))) | | | |

Figure 16.1: Wilcoxon Rank Sum Test—Formulas and Values

**Example 16.1.** (Exercises 16.1–16.4, pages 16.9–16.11 in Companion Chapter 16.) You are given growth rates for private consumption in 12 Asian and 13 Eastern European nations. Let

$$H_0 : \text{growth rates are the same}$$
$$H_a : \text{growth rates are higher in Asia}$$

Carry out a nonparametric test to judge whether the growth rate is higher in Asia than in Eastern Europe.

**Solution.** Figure 16.1 shows the Excel formulas required and the corresponding values taken when applied to the above data set.

1. Enter the labels "Asia," "Eastern Europe," "Population," "Combined," and "Rank" in cells A2:E2.

2. Enter the observations from Asia into cells A3:A14 and copy them to D3:D14. Enter the observations from Eastern Europe into cells B3:B15 and copy them to D15:D27. Enter the value 1 in C3:C14, the value 2 in C15:C27, and the values $\{1, \ldots, 25\}$ in E3:E27.

3. **Name** the ranges for Asia, Eastern Europe, Population, Combined, and Rank, because named ranges will be used in the calculations.

4. Next, we **Rank** the combined data set that has been copied into D3:D26 and carry the order of the ranking back to the corresponding population numbers in cells C3:C27, which identify the combined data. This is done as follows. Select C2:D27 and from the Menu Bar choose **Data – Sort** to bring up the **Sort** dialog box (Figure 16.2). Select the radio buttons **Ascending**

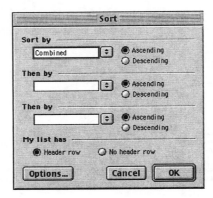

Figure 16.2: Sort Dialog Box

and **Header row** and make sure you **Sort by Combined**. You can select which column to use for sorting with the button to the right of the text entry box. Then click **OK**. This reorders the population labels in C3:C27 so that they move together with the corresponding entries in D3:D27. For example, the observation "-12.1" which was originally at the bottom of the combined data now is at the top with a combined rank of "1." Top half of Figure 16.1 shows the results after the sort.

5. Next, we carry out calculations for a large sample normal test for a mean. The bottom half of Figure 16.1 shows the details for lower, upper, and two-sided tests, with the formulas in column C and the values taken in column B. Because we are dealing with an upper test, the values are given only for the upper test, although formulas are shown for all three types of alternatives. Enter the labels as shown in column A of Figure 16.1 and

**Name** the corresponding ranges in the respective column B cells, in order to be able to refer to $n\_1$, $n\_2$, $W$, mu, sigma, $z$, lower$\_z$, upper$\_z$, and two$\_z$ by name in the formulas that are shown. The formulas to be entered in column B are presented in column C, and the values taken when the formulas are applied to this data (that is, what you will see in your workbook) are shown in the corresponding cells in column B. For instance, enter the formula from cell C32

$$= \text{SUMIF(Population, "= 1", Rank)}$$

into cell B32. This sums the cells under Rank whose corresponding Population is 1. In other words, the function adds the ranks corresponding to observations from Asia. The answer "198" appears in cell B32; this is the value of the Wilcoxon rank sum statistic. As noted, formulas for all three types of alternate hypotheses are provided in Figure 16.1, but the values are shown only for the particular alternative in this problem (upper test).

When using this template, remember to use only the appropriate cells for the problem at hand.

## Interpreting the Results

We read off

$$
\begin{aligned}
W &= 198 &&\text{(cell B33)} \\
\mu_W &= 156 &&\text{(cell B34)} \\
\sigma_W &= 18.385 &&\text{(cell B35)} \\
z \text{ statistic} &= 2.284 &&\text{(cell B36)}
\end{aligned}
$$

The alternate hypothesis is

$$H_a : \mu_1 - \mu_2 > 0$$

so that rows 41–44 are appropriate (rows 37–40 for a lower-tailed test and rows 45–48 for a two-tailed test). We find that

$$
\begin{aligned}
\text{upper critical 5\% value } z^* &= 1.645 &&\text{(cell B42)} \\
\text{decision rule} &= \text{"Reject H0"} &&\text{(cell B43)} \\
P\text{-value} &= 0.0112 &&\text{(cell B44)}
\end{aligned}
$$

We conclude that the data are significant at the nominal 5% level of significance.

## Continuity Correction for the Normal Approximation

A more accurate $P$-value is obtained by applying the continuity correction, which adjusts for the fact that a continuous distribution (the normal) is being used to approximate a discrete distribution $W$. If we use a correction of 0.5, then the upper-tailed test statistic becomes

$$z = \frac{W - 0.5 - \mu_W}{\sigma_W}$$

and this leads to

$$z = 2.257$$

and

$$P\text{-value} = 1 - \Phi(2.257) = 0.0120$$

## Ties

Theoretically, the assumption of a continuous distribution ensures that all $n_1 + n_2$ observed values will be different. In practice, ties are sometimes observed. The common practice is to average the ranks for the tied observations and carry on as above with a change in the standard deviation. Use

$$\sigma_W^2 = \frac{n_1 n_2}{12} \left( n_1 + n_2 + 1 - \frac{\sum_{i=1}^{G} t_i(t_i^2 - 1)}{(n_1 + n_2)(n_1 + n_2 - 1)} \right)$$

where $G$ is the number of tied groups and $t_i$ is the number of tied observations in the $i$th tied group. Unless $G$ is large, the adjustment in the formula for the variance makes little difference.

## 16.2  The Wilcoxon Signed Rank Test

The Wilcoxon signed rank test is a nonparametric version of the one-sample procedures based on the assumption of normal population discussed in Chapters 6 and 7. The key assumption is that the data arise from a population symmetric about its mean.

For this reason, one of its most useful applications is in a matched-pairs setting with $n$ pairs $(x_{1i}, x_{2i})$ of observations, where it is natural to assume that the populations from which the pairs are taken differ only by a shift in the mean (that is, the population distribution shapes are otherwise the same). The differences then satisfy the requirement of symmetry under the null hypothesis of equality of means.

### Description of the Matched-Pairs Procedure

The data consist of $n$ pairs $(x_{1i}, x_{2i})$ of observations. The $\{x_{1i}\}$ are a sample from a population with mean $\mu_1$ and the $\{x_{2i}\}$ are a sample from a population with mean $\mu_2$. The null hypothesis is

$$H_0 : \mu_1 - \mu_2 = 0$$

1. Form the absolute differences $|d_j|$, where $d_j = x_{1j} - x_{2j}$

2. Let $r_j$ be the rank of $|d_j|$ in the joint ranking of the $\{|d_j|\}$, from smallest to largest.

3. Form the sum of the positive signed ranks

$$W^+ = \sum r_j$$

where the sum is taken over all ranks $r_j$ for which the corresponding difference $d_j$ is positive.

The Wilcoxon signed rank procedure rejects $H_0$ if $W^+$ is beyond some reasonable value, in particular for values of $W^+$ that are too large or too small.

As with the Wilcoxon rank sum statistic $W$, there exist tables of the exact distribution of $W^+$, but we will base our procedure on the normal approximation. When $H_0$ is true the mean and the standard deviation of $W^+$ are given by

$$\mu_{W^+} = \frac{n(n+1)}{4}$$

$$\sigma_{W^+} = \sqrt{\frac{n(n+1)(2n+1)}{24}}$$

We then calculate

$$z = \frac{W^+ - \mu_{W^+}}{\sigma_{W^+}}$$

and for a fixed level $\alpha$ test:

$$\begin{array}{ll} \text{if } H_a : \mu_1 - \mu_2 > 0 & \text{reject } H_0 \text{ if } z > z^* \\ \text{if } H_a : \mu_1 - \mu_2 < 0 & \text{reject } H_0 \text{ if } z < -z^* \\ \text{if } H_a : \mu_1 - \mu_2 \neq 0 & \text{reject } H_0 \text{ if } |z| > z^* \end{array}$$

## The Wilcoxon Signed Rank Test in Practice

**Example 16.2.** (Examples 16.6–16.8, pages 16.21–16.24 in Companion Chapter 16.) Food products are often enriched with vitamins and other supplements. Does the level of a supplement decline over time, so that the user receives less than the manufacturer intended. Here are the data on the vitamin C levels (milligrams per 100 grams) in wheat soy blend, a flour-like product supplied by international aid programs mainly for feeding children. The same nine bags of blend were measured at the factory and five months later in Haiti.

| Bag | 1 | 2 | 3 | 4 | 5 | 6 | 7 | 8 | 9 |
|-----|----|----|----|----|----|----|----|----|----|
| Factory | 45 | 32 | 47 | 40 | 38 | 41 | 37 | 52 | 37 |
| Haiti | 38 | 40 | 35 | 38 | 34 | 35 | 38 | 38 | 40 |
| Difference | 7 | -8 | 12 | 2 | 4 | 6 | -1 | 14 | -3 |

We suspect that vitamin C levels are generally higher at the factory than they are five months later. We would like to test the hypotheses:

$H_0$ : vitamin C has the same distribution at both times

$H_a$ : vitamin C is systematically higher at the factory

Because this is a matched-pairs design, we base our inference on the differences and use a Wilcoxon signed rank test.

**Solution.**    Figure 16.3 shows the Excel formulas required and the corresponding values taken for the example data set. The calculations are similar to those in the previous section.

| | A | B | C | D | E | F | G | H | I | J | K |
|---|---|---|---|---|---|---|---|---|---|---|---|
| 1 | Wilcoxon Signed Rank Test - Matched Pairs | | | | | | | | | | |
| 2 | | Data after ranking | | | | | | Data before ranking | | | |
| 3 | Factory | Haiti | Diff | AbsDiff | Rank | | Factory | Haiti | Diff | AbsDiff | Rank |
| 4 | 37 | 38 | -1 | 1 | 1 | | 45 | 38 | 7 | 7 | 1 |
| 5 | 40 | 38 | 2 | 2 | 2 | | 32 | 40 | -8 | 8 | 2 |
| 6 | 37 | 40 | -3 | 3 | 3 | | 47 | 35 | 12 | 12 | 3 |
| 7 | 38 | 34 | 4 | 4 | 4 | | 40 | 38 | 2 | 2 | 4 |
| 8 | 41 | 35 | 6 | 6 | 5 | | 38 | 34 | 4 | 4 | 5 |
| 9 | 45 | 38 | 7 | 7 | 6 | | 41 | 35 | 6 | 6 | 6 |
| 10 | 32 | 40 | -8 | 8 | 7 | | 37 | 38 | -1 | 1 | 7 |
| 11 | 47 | 35 | 12 | 12 | 8 | | 52 | 38 | 14 | 14 | 8 |
| 12 | 52 | 38 | 14 | 14 | 9 | | 37 | 40 | -3 | 3 | 9 |
| 13 | | | | | | | | | | | |
| 14 | | | | | | | | | | | |
| 15 | alpha | 0.05 | | | | | | | | | |
| 16 | Calculations | | | | | | | | | | |
| 17 | n | 9 | =COUNT(Factory) | | | | | | | | |
| 18 | W+ | 34 | =SUMIF(Diff, ">0", Rank) | | | | | | | | |
| 19 | mu | 22.5 | =n*(n+1)/4 | | | | | | | | |
| 20 | sigma | 8.441 | =SQRT(n*(n+1)*(2*n+1)/24) | | | | | | | | |
| 21 | z | 1.362 | =(Wplus-mu)/sigma | | | | | | | | |
| 22 | Lower Test | | | | | | | | | | |
| 23 | lower_z | | =NORMSINV(alpha) | | | | | | | | |
| 24 | Decision | | =IF(z<lower_z,"Reject HO","Do Not Reject HO") | | | | | | | | |
| 25 | Pvalue | | =NORMSDIST(z) | | | | | | | | |
| 26 | Upper Test | | | | | | | | | | |
| 27 | upper_z | 1.645 | =-NORMSINV(alpha) | | | | | | | | |
| 28 | Decision | Do Not Reject | =IF(z>upper_z,"Reject HO","Do Not Reject HO") | | | | | | | | |
| 29 | Pvalue | 0.087 | =1-NORMSDIST(z) | | | | | | | | |
| 30 | Two-Sided Test | | | | | | | | | | |
| 31 | two_z | | =ABS(NORMSINV(alpha/2)) | | | | | | | | |
| 32 | Decision | | =IF(ABS(z)>two_z,"Reject HO","Do Not Reject HO") | | | | | | | | |
| 33 | Pvalue | | =2*(1-NORMSDIST(ABS(z))) | | | | | | | | |

Figure 16.3: Wilcoxon Signed Rank—Matched Pairs

1. Enter the labels "Factory," "Haiti," "Diff," "AbsDiff," and "Rank" in cells A3:E3.

2. Record the data for Factory in cells A4:A12 and the data for Haiti in cells B4:B12. In cells C4:C12 record the difference Factory−Haiti. In cells D4:D12 enter the absolute value of the differences using the Excel function **ABS()**. Finally enter the values $\{1, 2, \ldots, 9\}$ in E4:E12.

3. **Name** the ranges for the corresponding labels Factory, Haiti, Diff, AbsDiff, and Rank to include the respective cells in rows 4:12.

4. We are now going to order the data in A4:D12 according to the rank of the Absolute Difference. Then we will be able to sum the ranks corresponding to the positive members of these differences. We do this sorting as in the previous example except that we sort all the columns. (Actually, in truth, we only need to sort columns C and D, but then columns A and B get out of order, so we sort them also to maintain integrity of the data set.) This is done as follows. Select A3:D12 and from the Menu Bar choose **Data − Sort** to bring up the **Sort** dialog box. Select the radio buttons **Ascending** and for **Header row** and make sure you **Sort by** "AbsDiff." Then click **OK**. This reorders the rows according to the value in column AbsDiff. For example, the observation "-12.1" which was originally at the bottom of the combined data now is at the top with a combined rank of "1." The top half of Figure 16.1 shows "before" and "after" results. The top right is the data set as originally presented (see the table above). The top left half of Figure 16.1 shows the results after the sort.

5. The calculations required are shown in the lower portion of Figure 16.1. These are similar to those carried out for the rank sum test.

   The formulas are in column C while the the actual values taken by these formulas — the values that will appear on your workbook — are in column B. Refer to step 5 of the previous section for the analogous details, and remember to name all ranges used in the formulas shown.

## Interpreting the Results

We read off

$$
\begin{aligned}
W^+ &= 34 \qquad \text{(cell B18)} \\
\mu_{W+} &= 22.5 \qquad \text{(cell B19)} \\
\sigma_{W+} &= 8.441 \qquad \text{(cell B20)} \\
z \text{ statistic} &= 1.362 \qquad \text{(cell B21)}
\end{aligned}
$$

The alternate hypothesis requires an upper-tailed test for which

$$
P\text{-value} = 0.087 \qquad \text{(cell B29)}
$$

We conclude that the data are not significant.

## Continuity Correction for the Normal Approximation

As with the Wilcoxon rank sum test, a more accurate $P$-value is obtained with the continuity correction

$$
z = \frac{W^+ - 0.5 - \mu_{W+}}{\sigma_{W+}}
$$

and this leads to

$$z = 1.303$$

$$P\text{-value} = 1 - \Phi(1.303) = 0.0963$$

**Note:** These more accurate values are shown on page 16.23 in the Companion Chapter 16.

### Ties and Zero Values

If there are zeros among the differences $\{d_i\}$, discard them and use for $n$ the number of nonzero $\{d_i\}$. If there are any ties, then use the average rank for each set of tied observations and apply the procedure with variance

$$\sigma_{W+}^2 = \frac{1}{24} \left( n(n+1)(2n+1) - \frac{\sum_{i=1}^{G} t_i(t_i^2 - 1)}{2} \right)$$

where $G$ is the number of tied groups and $t_i$ are the number of tied observations in the $i$th tied group.

## 16.3    The Kruskal-Wallis Test

In this section, we generalize the Wilcoxon rank sum test to situations involving independent samples from $I$ populations when the assumptions required for validity of the one-way ANOVA in Chapter 12 cannot be substantiated.

The data consist of $N = \sum_{i=1}^{I} n_i$ observations with $n_i \geq 1$ observations $\{x_{ij} : 1 \leq j \leq n_i\}$ taken from population $i$. The assumption replacing normality is

$$x_{ij} = \mu_i + \varepsilon_{ij} \quad 1 \leq i \leq I, \ 1 \leq j \leq n_i$$

where the errors $\{\varepsilon_{ij}\}$ are mutually independent with mean 0 and have the same *continuous* distribution. If we let $F(x)$ be the cumulative distribution function (c.d.f.) of a generic error term, this assumption is tantamount to $F_i(x)$ being the c.d.f. of population $i$, where

$$F_i(x) \equiv F(x - \mu_i) \quad 1 \leq i \leq I$$

The significance test is

$$H_0 : \mu_1 = \mu_2 = \cdots = \mu_I$$
$$H_a : \text{not all of the } \mu_i \text{ are equal}$$

The procedure generalizing the rank sum test is called the Kruskal-Wallis test.

### Description of the Procedure

1. **Rank** all the observations jointly from smallest to largest.

2. Let $r_{ij}$ be the rank of observation $x_{ij}$.

3. Set

$$R_i = \sum_{j=1}^{n_i} r_{ij}$$

which is the sum of the ranks associated with sample $i$.

Denote by

$$\bar{R}_i = \frac{1}{n_i} R_i$$

the average rank in sample $i$. If $H_0$ is true, then by symmetry the mean of any rank $r_{ij}$ is $E(r_{ij}) = \frac{N+1}{2}$, which is the average of the integers $\{1, 2, \ldots, N\}$, and therefore $E[\bar{R}_i] = E\left[\frac{1}{n_i} R_i\right] = E\left[\frac{1}{n_i} \sum_{j=1}^{n_i} r_{ij}\right] = \frac{N+1}{2}$. Thus, we would expect the ranks to be uniformly intermingled among the $I$ samples. But if $H_0$ is false, then some samples will tend to have many small ranks, while others will have many large ranks. Just as in ANOVA, we take the sum of squares of the differences between the average rank $\bar{R}_i$ of each sample and the overall average $\frac{N+1}{2}$ by computing

$$\frac{12}{N(N+1)} \sum_{i=1}^{I} n_i \left(\bar{R}_i - \frac{N+1}{2}\right)^2$$

which can be expressed equivalently as

$$H = \frac{12}{N(N+1)} \sum_{i=1}^{I} \frac{R_i^2}{n_i} - 3(N+1)$$

and called the Kruskal-Wallis statistic. We then reject $H_0$ for "large" values of $H$.

Tables of critical values exist for small values of the $\{n_i\}$, but it is customary to use a normal approximation, which provides an approximate sampling distribution:

*H is approximately chi-square with $I - 1$ degrees of freedom.*

Therefore the test is

$$\text{Reject } H_0 \text{ if } H > \chi^2$$

where $\chi^2$ is the upper critical $\alpha$ value of a chi-square distribution on $I - 1$ degrees of freedom.

## The Kruskal-Wallis Test in Practice

**Example 16.3.** (Exercise 16.33, page 16.36 in Companion Chapter 16.) Decay of polyester fabric. Here are the breaking strengths (in pounds) of grips of polyster fabric buried in the ground for several lengths of time.

| Time | Breaking strength | | | | |
|---|---|---|---|---|---|
| 2 weeks | 118 | 126 | 126 | 120 | 129 |
| 4 weeks | 130 | 120 | 114 | 126 | 128 |
| 8 weeks | 122 | 136 | 128 | 146 | 140 |
| 12 weeks | 124 | 98 | 110 | 140 | 110 |

(a) Find the standard deviations of the 4 samples.

(b) Find the medians.

(c) Carry out the Kruskal-Wallis test on the medians.

**Solution.** Figure 16.4 shows the Excel formulas required and the corresponding values taken when applied to the example data set. Enter the data and labels into the block of cells A3:D8.

(a) and (b) Use the Excel functions MEDIAN() and STDEV() to find the medians and standard deviations. These are shown in A10:D11.

| | Median | Std Dev |
|---|---|---|
| 2 weeks | 126 | 4.604 |
| 4 weeks | 126 | 6.542 |
| 8 weeks | 136 | 9.529 |
| 12 weeks | 110 | 16.087 |

The sample standard deviations do not satisfy our rule of thumb that, for safe use of ANOVA, the largest should not exceed twice the smallest, so we use a nonparametric test.

(c) To use the Kruskal-Wallis test, first combine all the observations into one column by copying A4:A8 into F4:F8, followed by B4:B8 into G9:G13, and so on. Then indicate which population the combined data come from in column F. Thus, enter the value "1" into cells F4:F8, the value "2" into cells F9:F13, and so on. Finally, enter the values $\{1, \ldots, 20\}$ in H4:H23. Label columns F, G, and H as "Pop," "Combined," and "Rank," respectively. Now proceed as in Example 16.1 by sorting columns F and G according to the column Combined, by selecting J3:K23. Then from the Menu Bar choose **Data – Sort** and sort by Combined. In Figure 16.4 columns F and G show the sorted data while columns J and K show what F and G looked like prior to the sort.

As we have already described in detail the construction of the two earlier nonparametric procedures, we leave as an exercise the application of this workbook. Beginning in row 13, column C shows the formulas to be entered into the adjacent cells of column B, where the numerical evaluation of the formulas is shown.

| | A | B | C | D | E | F | G | H | I | J | K |
|---|---|---|---|---|---|---|---|---|---|---|---|
| 1 | | | Kruskal-Wallis Test | | | | | | | | |
| 2 | | | | | | | | | | | |
| 3 | 2 weeks | 4 weeks | 8 weeks | 12 weeks | | Pop | Combined | Rank | | Pop | Combined |
| 4 | 118 | 130 | 122 | 124 | | 4 | 98 | 1 | | 1 | 118 |
| 5 | 126 | 120 | 136 | 98 | | 4 | 110 | 2 | | 1 | 126 |
| 6 | 126 | 114 | 128 | 110 | | 4 | 110 | 3 | | 1 | 126 |
| 7 | 120 | 126 | 146 | 140 | | 2 | 114 | 4 | | 1 | 120 |
| 8 | 129 | 128 | 140 | 110 | | 1 | 118 | 5 | | 1 | 129 |
| 9 | | | | | | 1 | 120 | 6 | | 2 | 130 |
| 10 | 126 | 126 | 136 | 110 | | 2 | 120 | 7 | | 2 | 120 |
| 11 | 4.604 | 6.542 | 9.529 | 16.087 | | 3 | 122 | 8 | | 2 | 114 |
| 12 | | | | | | 4 | 124 | 9 | | 2 | 126 |
| 13 | R_1 | 47 | =SUMIF(Pop, "=1", Rank) | | | 1 | 126 | 10 | | 2 | 128 |
| 14 | R_2 | 52 | =SUMIF(Pop, "=2", Rank) | | | 1 | 126 | 11 | | 3 | 122 |
| 15 | R_3 | 77 | =SUMIF(Pop, "=3", Rank) | | | 2 | 126 | 12 | | 3 | 136 |
| 16 | R_4 | 34 | =SUMIF(Pop, "=4", Rank) | | | 2 | 128 | 13 | | 3 | 128 |
| 17 | | | | | | 3 | 128 | 14 | | 3 | 146 |
| 18 | n_1 | 5 | =COUNT(A4:A8) | | | 1 | 129 | 15 | | 3 | 140 |
| 19 | n_2 | 5 | =COUNT(B4:B8) | | | 2 | 130 | 16 | | 4 | 124 |
| 20 | n_3 | 5 | =COUNT(C4:C8) | | | 3 | 136 | 17 | | 4 | 98 |
| 21 | n_4 | 5 | =COUNT(D4:D8) | | | 3 | 140 | 18 | | 4 | 110 |
| 22 | N | 20 | =n_1+n_2+n_3+n_4 | | | 4 | 140 | 19 | | 4 | 140 |
| 23 | | | | | | 3 | 146 | 20 | | 4 | 110 |
| 24 | | 441.8 | =R_1^2/n_1 | | | | | | | | |
| 25 | | 540.8 | =R_2^2/n_2 | | | | | | | | |
| 26 | | 1185.8 | =R_3^2/n_3 | | | | | | | | |
| 27 | | 231.2 | =R_4^2/n_4 | | | | | | | | |
| 28 | | | | | | | | | | | |
| 29 | H | 5.56 | =(12/(N*(N+1)))*SUM(B24:B27) - 3*(N+1) | | | | | | | | |
| 30 | Critical 5% | 7.815 | =CHIINV(0.05,3) | | | | | | | | |
| 31 | P-value: | 1.35E-01 | =CHIDIST(H,3) | | | | | | | | |

Figure 16.4: Kruskal-Wallis Test—Formulas and Values

## Interpreting the Results

We read off

$$
\begin{aligned}
H &= 5.56 \quad \text{(cell B29)} \\
\chi^2 &= 7.815 \quad \text{(cell B30)} \\
P\text{-value} &= 0.135 \quad \text{(cell B31)}
\end{aligned}
$$

The data are not significant at the 5% level, meaning that there is no convincing evidence that length of time in the ground affects the breaking strength.

## Ties

We have ignored ties in the above calculation. For a more accurate calculation, give all tied values their average ranks. For instance, the second and third smallest values in the combined list are tied at the value 110. We give each of them the rank 2.5, which is half of 2+3. After assigning the averages to tied ranks, we then replace $H$ with

$$H' = \frac{H}{1 - \sum_{i=1}^{G} \frac{t_i(t_i^3 - 1)}{N^3 - N}}$$

where $G$ is the number of tied groups and $t_i$ are the number of tied observations in the $i$th tied group. A separate calculation shows that this gives us $H' = 5.60$ and a $P$-value of 0.128.

# Chapter 17

# Logistic Regression

Excel does not possess a built-in logistic regression tool. Logistic regression is a highly specialized topic, best used under the guidance of a statistician. However, it is still possible within Excel to obtain estimates of the parameters in simple logistic regression with little effort, using weighted least squares. This illustrates the power of spreadsheet operations.

## 17.1 The Logistic Regression Model

Recall from the discussion in Section 2.5 and Chapter 10 that regression refers to fitting models for the mean value of a response as a function of an explanatory variable. For $n$ pairs of observations $(x_i, y_i)$ the simple linear regression model is

$$y_i = \beta_0 + \beta_1 x_i + \varepsilon_i$$

with the errors $\{\varepsilon_i\}$ assumed independent $N(0, \sigma)$ and the regression function

$$\mu_y = \beta_0 + \beta_1 x_i$$

representing the mean of $y_i$ as a function of $x$.

It is possible to fit models *other* than a straight line. If you examine Figure 2.14, you will observe that Excel can fit

$$\mu_y = a + b \log x \quad \text{(logarithmic)}$$
$$\mu_y = ax^b \quad \text{(power)}$$
$$\mu_y = ae^{bx} \quad \text{(exponential)}$$

as well as polynomial and moving average models. These nonlinear models are fitted using transformations that linearize the regression curve.

In some circumstances, the response variable is discrete, not continuous. An example of a discrete variable might be the number of cases of skin cancer in a metropolitan area. A special case of a discrete variable is a binary response, say $\{0, 1\}$, that leads to the logistic regression model. We restrict attention to binary response in the remainder of this chapter.

## Binomial Distributions and Odds

First, we will identify some major differences between regression with binary responses and regression with continuous responses. Assume, as is customary, that the errors have mean 0, which endows the regression curve

$$\mu_y = \beta_0 + \beta_1 x$$

with the usual meaning as the mean response. In the Bernoulli case, $y_i$ can take only two values so that

$$\mu_y = P[y_i = 1] = \beta_0 + \beta_1 x_i = p_i$$

and the variance of the error term is therefore the variance of a Bernoulli random variable, $p_i(1 - p_i)$. Note:

- Errors cannot be normal.
- Errors have nonconstant variance.
- The mean response is constrained to lie within the interval [0,1] because it represents a probability.

Because of these differences, ordinary regression methodology is not appropriate.

The explanatory variables are $(x_1, x_2, \ldots, x_c)$ with $n_i$ observations at the value $x_i$. The number of "successes" at $x_i$ is denoted by $s_i$, which is a binomial $\text{Bin}(n_i, p_i)$ random variable on $n_i$ trials and success probability $p_i \equiv p(x_i)$, which is a function of $x_i$. It is the function $p(x_i)$ that is of interest.

We have dealt with binary responses on two occasions in this text; in Chapter 8 we considered $c = 2$, while in Chapter 9 we dealt with $c \geq 2$. Here we also consider $c \geq 2$ but impose a model (the regression curve) relating all the probabilities $p_i$ as a function of $x_i$.

A scatterplot of such data shows binary observations having $y$ values of 0 and 1; the interpretation of the regression curve as somehow passing near the data is lost because of the categorical (binary) nature of the response. However, the interpretation is partially regained if a scatterplot is made of the sample proportions $\hat{p}_i = \frac{s_i}{n_i}$ on the $y$-axis against their corresponding $x_i$ on the $x$-axis.

We may then consider fitting a curve through the $(x_i, \hat{p}_i)$ pairs. In order to facilitate this approach, statisticians have introduced the logit transformation based on the odds. Define the odds ratio

$$\text{ODDS} = \frac{p}{1 - p}$$

and then take the natural logarithm of the odds to define the logit function:

$$\text{logit}(p) = \log(\text{ODDS}) = \log\left(\frac{p}{1 - p}\right)$$

There is a mathematical reason for using ODDS as the *natural* parameter that comes from the factorization of the likelihood function. These are advanced details, beyond the scope of this presentation.

## The Statistical Model

The **statistical model for logistic regression** posits a linear form

$$\text{logit}(p) = \beta_0 + \beta_1 x$$

that is equivalent to

$$p \equiv p(x) = \frac{1}{1 + e^{-(\beta_0 + \beta_1 x)}}$$

in terms of the original probability $p$ (Figure 15.4 of the Student CD-ROM).

In the text, estimates for the parameters $\beta_0, \beta_1$ are presented based, on the output from a specialized statistics package called SAS. Although Excel does not provide a logistic regression output, we can obtain estimates using the **spreadsheet** features of Excel and **weighted least-squares**.

## Weighted Least-Squares

The least-squares criterion minimizes the sum of the squared residuals

$$\sum_{i=1}^{n} e_i^2 = \sum_{i=1}^{n} (y_i - \hat{y}_i)^2$$

where $\{y_i\}$ are the observed values and $\{\hat{y}_i\}$ are the fitted values where $\hat{y}_i = b_0 + b_1 x_i$ . When the errors do not have a constant variance, it is more **efficient** (in the sense of producing estimates with smaller variance) to weight the residuals. It is intuitively reasonable to give a residual more weight (for accuracy) if its corresponding variance is smaller.

Let $w_i$ be the weight assigned to an observation at $x_i$. Then $w_i$ is inversely proportional to the variance $\sigma_i^2$ of the error $\varepsilon_i$ and the criterion for weighted least-squares is

$$\text{minimize} \quad \sum_{i=1}^{n} w_i (y_i - b_0 - b_1 x_i)^2$$

The solution is obtained (as in Chapter 2) by differentiating with respect to $b_0, b_1$. We find

$$b_1 = \frac{\sum w_i x_i y_i - \frac{(\sum w_i x_i)(\sum w_i y_i)}{\sum w_i}}{\sum w_i x_i^2 - \sum w_i x_i}$$

$$b_0 = \frac{\sum w_i y_i - b_1 \sum w_i x_i}{\sum w_i}$$

When $w_i \equiv 1$ for all $i$, these two equations reduce to the equations obtained for finding the ordinary least-squares estimates.

These equations represent the quintessential spreadsheet operations, columns of numbers that are added and whose totals are then algebraically manipulated.

Therefore, they are ideal for spreadsheet logic, and *it is fitting in this chapter, dealing with a sophisticated statistical model for which Excel does not provide a built-in tool, that we can obtain a solution by setting up our own columns of variables.*

The variable $y_i$, which is appropriate in this setting, is

$$y_i = \log \frac{\hat{p}_i}{1 - \hat{p}_i}$$

and the weights are approximately $w_i = n_i p_i (1 - p_i)$. Because the $\{p_i\}$ are unknown, we employ instead

$$w_i = n_i \hat{p}_i (1 - \hat{p}_i)$$

We organize our workbook with columns for

$$x_i, y_i, w_i, w_i x_i, w_i y_i, w_i x_i^2, w_i y_i, w_i x_i y_i$$

which are then used to determine $b_0, b_1$ (Figure 17.1).

## 17.2   Inference for Logistic Regression

**Example 17.1.** (Examples 17.7, page 17.14 in Companion Chapter 17.) An experiment was designed to examine how well the insecticide rotenone kills aphids that feed on the chrysanthemum plant called *macrosiphoniella sanborni*. The explanatory variable is concentration (in log of mg/l) of the insecticide. About 50 aphids were exposed to each of five concentrations. Each insect was either killed or not killed. The response variable for logistic regression is the log odds of the proportion killed. Although logistic regression uses the log odds of the proportion killed, Excel requires that we enter just the concentration, the number of insects, and the number killed. Here are the data, along with the results of some calculations:

| Concentration $x$ (log scale) | Number of insects | Number killed | Proportion killed $\hat{p}$ | log ods |
|---|---|---|---|---|
| 0.96 | 50 | 6 | 0.1200 | -1.9924 |
| 1.33 | 48 | 16 | 0.3333 | -0.6931 |
| 1.63 | 46 | 24 | 0.5217 | 0.0870 |
| 2.04 | 49 | 42 | 0.8571 | 1.7918 |
| 2.32 | 50 | 44 | 0.8800 | 1.9924 |

Fit a logistic model

$$p_i = \frac{1}{1 + e^{-(\beta_0 + \beta_1 x_i)}}$$

to the probability that an aphid will be killed as a function of the concentration $x_i$.

| | A | B | C | D | E | F | G | H | I | J | K |
|---|---|---|---|---|---|---|---|---|---|---|---|
| 1 | | | | | **Logistic Regression by Weighted Least Squares** | | | | | | |
| 2 | | | | | | | | | | | |
| 3 | | | | | logit | weight | | | calculations | | |
| 4 | x | n | s | p | y | w | w*x | w*y | w*x*y | w*x^2 | w*y^2 |
| 5 | 0.96 | 50 | 6 | 0.120 | -1.992 | 5.280 | 5.069 | -10.520 | -10.099 | 4.866 | 20.960 |
| 6 | 1.33 | 48 | 16 | 0.333 | -0.693 | 10.667 | 14.187 | -7.394 | -9.833 | 18.868 | 5.125 |
| 7 | 1.63 | 46 | 24 | 0.522 | 0.087 | 11.478 | 18.710 | 0.999 | 1.628 | 30.497 | 0.087 |
| 8 | 2.04 | 49 | 42 | 0.857 | 1.792 | 6.000 | 12.240 | 10.751 | 21.931 | 24.970 | 19.262 |
| 9 | 2.32 | 50 | 44 | 0.880 | 1.992 | 5.280 | 12.250 | 10.520 | 24.406 | 28.419 | 20.960 |
| 10 | | | sums | | | 38.705 | 62.455 | 4.356 | 28.033 | 107.620 | 66.395 |
| 11 | | | | | | | | | | | |
| 12 | | | | =s/n | =LN(p/(1-p)) | =n*p*(1-p) | =w*x | =w*y | =w*x*y | =w*x^2 | =w*y^2 |
| 13 | | | | | | | | | | | |
| 14 | b1= | 3.070 | =(I10-G10*H10/F10)/(J10-G10^2/F10) | | | | | | | | |
| 15 | b0= | -4.841 | =(H10-B14*G10)/F10 | | | | | | | | |

Figure 17.1: Weighted Least-Squares Estimates for Logistic Regression

**Solution.** Figure 17.1 shows how we have set up the workbook. Use **Named Ranges** for the labels and variables shown in row 4. The formulas required for calculating the columns are shown in cells D12:K12. The weighted least-squares estimates are in B14:B15, and in C14:C15 we have given the formulas to be entered into B14:B15. (Your workbook will not have C14:C15.)

We find the weighted least-squares estimates to be

$$b_1 = 3.070$$
$$b_0 = -4.841$$

These compare favorably with the SAS output presented in the text, which shows that the student can manage some sophisticated analyses using spreadsheet operations.

$$b_1 = 3.10$$
$$b_0 = -4.89$$

**Example 17.2.** (Examples 17.1–17.4, pages 17.5–17.10 in Companion Chapter 17.) In Chapter 8 we used sample proportions to estimate population proportions in a survey on binge drinking. Adapt the workbook developed in the previous example to show that the weighted least-squares estimates are

$$b_1 = 0.362$$
$$b_0 = -1.587$$

These are identical to those in your text on page 17.9 Your results should look like those in Figure 17.2.

**Example 17.3.** Adapt the workbook shown in Figure 17.1 and the formula $= \text{CRITBINOM}(n, p, \text{RAND}())$ to simulate data for a logistic regression.

| | A | B | C | D | E | F | G | H | I | J | K |
|---|---|---|---|---|---|---|---|---|---|---|---|
| 1 | | | | | \multicolumn{7}{c}{**Logistic Regression by Weighted Least Squares**} | | | | | |
| 2 | | | | | | | | | | | |
| 3 | | | | | logit | weight | | calculations | | | |
| 4 | x | n | s | p | y | w | w*x | w*y | w*x*y | w*x^2 | w*y^2 |
| 5 | 0 | 9916 | 1684 | 0.170 | -1.587 | 1398.012 | 0.000 | -2218.445 | 0.000 | 0.000 | 3520.356 |
| 6 | 1 | 7180 | 1630 | 0.227 | -1.225 | 1259.958 | 1259.958 | -1543.723 | -1543.723 | 1259.958 | 1891.398 |
| 7 | | | sums | | | | 2657.970 | 1259.958 | -3762.169 | -1543.723 | 1259.958 | 5411.753 |
| 8 | | | | | | | | | | | |
| 9 | | | | | =s/n | =LN(p/(1-p)) | =n*p*(1-p) | =w*x | =w*y | =w*x*y | =w*x^2 | =w*y^2 |
| 10 | | | | | | | | | | | |
| 11 | b1= | 0.362 | = (I7-G7*H7/F7)/(J7-G7^2/F7) | | | | | | | | |
| 12 | b0= | -1.587 | = (H7-B11*G7)/F7 | | | | | | | | |

Figure 17.2: Binge Drinking

Choose the same values $\{x_i\}$ and $\{n_i\}$ as in Example 17.1. Set the parameters to be

$$\beta_1 \;=\; -5.00$$
$$\beta_0 \;=\; \phantom{-}3.00$$

and obtain estimates $b_1, b_0$.

Next, generate repeated samples and construct histograms of the values for $b_0$ and $b_1$. Compare them with the true values. What can you conclude about the mean, bias, standard error, and confidence intervals? Construct scatterplots of the simulated values of $(b_0, b_1)$.

# Chapter 18

# Bootstrap Methods and Permutation Tests

## 18.1  Why Resampling?

Bootstrap methods are based on resampling from data and were first introduced in 1979 for estimating the standard error of the estimate of a parameter. The bootstrap is now used in more general situations including confidence intervals, significance testing, and regression, for instance. The bootstrap is based on the empirical distribution function and is best understood from the "plug-in" principle. This principle estimates parameters by applying to the data the same function which is applied to the population to obtain the parameter of interest. The simplest illustration of this principle is the sample mean. The sample mean is the average of the data and the population mean is the average of the population.

The bootstrap is best carried out with specialized software that does simulation well and quickly. Excel does not possess a built-in bootstrap and it is not particularly suited to large-scale simulations without additional macros. Nonetheless we can demonstrate the principles of the bootstrap with spreadsheet and graphing operations for which Excel is well-suited.

## 18.2  Introduction to Bootstrapping

### The Procedure

1. Resample with replacement $N$ times from the data set $\{x_1, x_2, \ldots, x_n\}$.

2. Calculate the statistic of interest for each sample. This gives the **bootstrap distribution** which may be visualized as a histogram representing the values of this statistic for all $N$ samples.

3. Use the bootstrap distribution to learn about the sampling properties of the statistic.

**Example 18.1.**   (Exercise 18.1, page 18.10 in Companion Chapter 18.) **Bootstrap a small data set**. The following is a small random sample of the Verizon data in the data file *verizon.dat* on the CD.

| 3.12 | 0.00 | 1.57 | 19.67 | 0.22 | 2.20 |

(a) Sample with replacement 20 times and create 20 samples each of size six.
(b) Calculate the sample mean for each of the resamples.
(c) Make a histogram of the 20 resamples.  This is the bootstrap distribution.
(d) Calculate the bootstrap standard error.

**Solution.**

| | A | B | C | D | E | F | G | H | I | J | K |
|---|---|---|---|---|---|---|---|---|---|---|---|
| 1 | **Bootstrap** | | | Bootstrap | | | | | | | Sample |
| 2 | | | | Sample | | | | | | | Mean |
| 3 | 3.12 | 0.1667 | | 1 | 3.12 | 19.67 | 0.00 | 0.22 | 3.12 | 19.67 | 7.63 |
| 4 | 0.00 | 0.1667 | | 2 | 1.57 | 19.67 | 1.57 | 0.00 | 1.57 | 0.22 | 4.10 |
| 5 | 1.57 | 0.1667 | | 3 | 0.00 | 0.00 | 19.67 | 0.22 | 2.20 | 0.22 | 3.72 |
| 6 | 19.67 | 0.1667 | | 4 | 1.57 | 19.67 | 19.67 | 0.00 | 0.22 | 0.22 | 6.89 |
| 7 | 0.22 | 0.1667 | | 5 | 2.20 | 0.00 | 1.57 | 0.22 | 0.22 | 0.00 | 0.70 |
| 8 | 2.20 | 0.1667 | | 6 | 3.12 | 0.22 | 3.12 | 19.67 | 0.00 | 19.67 | 7.63 |
| 9 | 4.46 | | | 7 | 19.67 | 1.57 | 0.00 | 19.67 | 0.22 | 0.00 | 6.86 |
| 10 | 7.54 | | | 8 | 2.20 | 2.20 | 1.57 | 0.22 | 3.12 | 19.67 | 4.83 |
| 11 | | | | 9 | 19.67 | 0.22 | 2.20 | 1.57 | 0.22 | 19.67 | 7.26 |
| 12 | | | | 10 | 3.12 | 1.57 | 3.12 | 19.67 | 1.57 | 3.12 | 5.36 |
| 13 | | | | 11 | 0.22 | 3.12 | 3.12 | 0.00 | 2.20 | 19.67 | 4.72 |
| 14 | | | | 12 | 1.57 | 1.57 | 3.12 | 1.57 | 0.00 | 1.57 | 1.57 |
| 15 | | | | 13 | 3.12 | 2.20 | 0.22 | 1.57 | 19.67 | 3.12 | 4.98 |
| 16 | | | | 14 | 2.20 | 3.12 | 1.57 | 0.22 | 0.22 | 0.22 | 1.26 |
| 17 | | | | 15 | 0.00 | 0.22 | 3.12 | 2.20 | 3.12 | 3.12 | 1.96 |
| 18 | | | | 16 | 2.20 | 0.22 | 3.12 | 0.00 | 19.67 | 3.12 | 4.72 |
| 19 | | | | 17 | 2.20 | 1.57 | 2.20 | 1.57 | 3.12 | 0.22 | 1.81 |
| 20 | | | | 18 | 3.12 | 0.22 | 0.00 | 3.12 | 0.00 | 1.57 | 1.34 |
| 21 | | | | 19 | 0.00 | 0.22 | 0.22 | 0.00 | 0.00 | 0.22 | 0.11 |
| 22 | | | | 20 | 19.67 | 2.20 | 0.00 | 3.12 | 19.67 | 2.20 | 7.81 |
| 23 | | | | | | | | | | | 4.26 |
| 24 | | | | | | | | | | | 2.58 |

Figure 18.1: Bootstrap Data and Samples

Step 1.  Enter the six data points into a column of a worksheet (as in cells A3:A8 in Figure 18.1).  In the adjacent column to the right enter the sampling probabilities ($\frac{1}{6}$) that give equal weight to each datum.

Step 2. From the Menu Bar chooses **Tools – Data Analysis . . . Random Number Generation** to bring up the **Random Number Generation** dialog box and complete as in Figure 18.2. We have chosen to have 20 rows rather than 20 columns for display purposes.  The 20 samples of size six appear in rows in E3:J22.

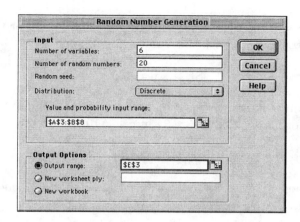

Figure 18.2: Random Number Generation Dialog Box

Step 3. Now calculate the means of each bootstrap sample by entering `= AVERAGE(E3:J3)` in cell K3, and filling down to cell K22.

Step 4. Using the **Histogram Tool** construct a histogram of these 20 bootstrap samples (Figure 18.3). This is the bootstrap distribution. It is clearly far from normal because the sample size (six) is small and because the original sample shows wide variability.

Step 5. To find the bootstrap standard error enter the formula `= STDEV(K3:K22)` on the worksheet (cell K24 in Figure 18.1). We get

$$SE_{boot} = 2.58$$

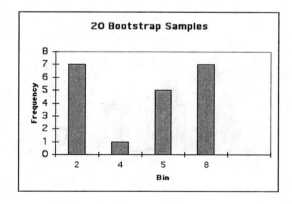

Figure 18.3: 20 Bootstrap Samples

**Note:** Let's compare the bootstrap estimated standard error or the sample mean $\bar{x}$ based on six observations with the usual estimate

$$\text{SE}_{\text{usual}} = s/\sqrt{n}$$

where $s$ is the sample standard deviation of the original sample. This can be found by entering the formula =STDEV(A3:A8) in cell A10 to find that $s = 7.54$ so that

$$\text{SE}_{\text{usual}} = 7.54/\sqrt{6} = 3.08$$

A better agreement can be made if we recognize that the bootstrap estimate assumes that that sample of size six is the actual population from which the repeated samples are drawn. Therefore the value 2.58 needs to be compared with the standard deviation of the original sample when it is viewed as the entire population. This mandates division by $n$ not by $n-1$ in the formula for $s$. So we want =STDEVP(A3:A8) for calculating $s$, and this differs from =STDEV(A3:A8) by a factor $\sqrt{\frac{n-1}{n}}$. As a result we obtain

$$\text{SE}_{\text{n}} = 6.88/\sqrt{6} = 2.81$$

giving $\frac{6.88}{\sqrt{6}} = 2.81$ which is in closer agreement with 2.58

We have also shown in Figure 18.4 the histogram based on 1000 bootstrap samples from the original sample of size six. The bootstrap distribution is far from normal because of the small sample size of the data set.

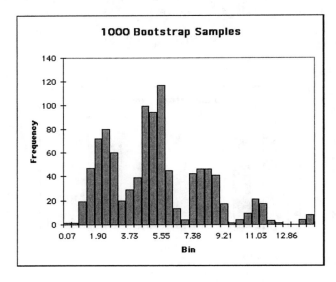

Figure 18.4: 1000 Bootstrap Samples

## 18.3 Bootstrap Distributions and Standard Errors

The bootstrap can be applied to other parameters than the mean. Moreover it can be used to find approximate confidence intervals. The next example illustrates how to find the bootstrap confidence for a mean and compares it with the usual interval based on the $t$ distribution.

| | A | B | C | D | E | F | G | AZ | BA | BB | BC |
|---|---|---|---|---|---|---|---|---|---|---|---|
| 1 | Bootstrap Confidence Interval for Mean | | | | 3.11 | 28.08 | 10.81 | 15.23 | 28.06 | 82.70 | 33.14 |
| 2 | | | | | 17.39 | 59.07 | 23.04 | 19.54 | 46.69 | 61.22 | 37.18 |
| 3 | 3.11 | 0.02 | | | 38.64 | 85.76 | 3.11 | 24.47 | 24.47 | 44.67 | 35.63 |
| 4 | 18.26 | 0.02 | | | 61.22 | 20.59 | 48.65 | 45.40 | 25.13 | 32.03 | 29.81 |
| 5 | 24.58 | 0.02 | =STDEV(BC1:BC1000) | | 48.65 | 41.02 | 22.22 | 28.38 | 20.16 | 52.75 | 35.26 |
| 6 | 36.37 | 0.02 | 3.001 | | 22.22 | 13.78 | 48.65 | 85.76 | 61.22 | 85.76 | 36.85 |
| 7 | 50.39 | 0.02 | | | 19.54 | 36.37 | 17.00 | 45.40 | 18.26 | 3.11 | 32.02 |
| 8 | 8.88 | 0.02 | =STDEV(A3:A52)/SQRT(50) | | 42.97 | 22.22 | 22.22 | 42.97 | 28.06 | 28.38 | 33.33 |
| 9 | 18.43 | 0.02 | 3.069 | | 32.03 | 24.47 | 44.08 | 18.26 | 27.65 | 3.11 | 35.19 |
| 10 | 25.13 | 0.02 | | | 17.00 | 93.34 | 52.75 | 61.22 | 41.02 | 44.08 | 35.16 |
| 11 | 38.64 | 0.02 | | | 34.98 | 38.64 | 26.26 | 24.47 | 26.26 | 3.11 | 37.07 |
| 12 | 52.75 | 0.02 | | | 44.08 | 61.22 | 39.16 | 44.08 | 24.58 | 85.76 | 32.67 |
| 13 | 9.26 | 0.02 | | | 50.39 | 17.00 | 24.47 | 24.47 | 15.62 | 20.59 | 37.03 |
| 14 | 19.27 | 0.02 | | | 52.75 | 38.64 | 28.06 | 18.43 | 54.80 | 3.11 | 36.10 |
| 15 | 26.24 | 0.02 | | | 28.38 | 26.26 | 15.62 | 12.69 | 24.47 | 26.24 | 36.28 |
| 16 | 39.16 | 0.02 | | | 82.70 | 59.07 | 41.02 | 24.47 | 46.69 | 44.08 | 34.90 |
| 17 | 54.80 | 0.02 | | | 50.39 | 42.97 | 23.04 | 24.47 | 9.26 | 9.26 | 29.86 |
| 18 | 10.81 | 0.02 | | | 12.69 | 32.03 | 19.54 | 10.81 | 42.97 | 17.00 | 35.12 |
| 19 | 19.50 | 0.02 | | | 85.76 | 17.00 | 20.16 | 26.26 | 24.58 | 25.13 | 31.75 |
| 20 | 26.26 | 0.02 | | | 61.22 | 82.70 | 93.34 | 28.06 | 82.70 | 70.32 | 38.23 |
| 21 | 41.02 | 0.02 | | | 28.38 | 39.16 | 45.40 | 8.88 | 18.43 | 86.37 | 29.83 |
| 22 | 59.07 | 0.02 | | | 19.50 | 44.67 | 54.80 | 24.58 | 17.39 | 23.04 | 34.80 |
| 23 | 12.69 | 0.02 | | | 36.37 | 26.24 | 70.32 | 17.39 | 39.16 | 15.62 | 33.23 |
| 24 | 19.54 | 0.02 | | | 54.80 | 24.47 | 28.06 | 22.22 | 32.03 | 8.88 | 33.40 |
| 25 | 27.65 | 0.02 | | | 17.39 | 28.38 | 26.24 | 39.16 | 20.59 | 9.26 | 29.07 |
| 26 | 42.97 | 0.02 | | | 24.47 | 23.04 | 39.16 | 28.06 | 10.81 | 24.58 | 33.77 |
| 27 | 61.22 | 0.02 | | | 27.65 | 28.38 | 10.81 | 41.02 | 17.00 | 45.40 | 31.15 |
| 28 | 13.78 | 0.02 | | | 28.38 | 23.04 | 46.69 | 19.54 | 52.75 | 54.80 | 35.24 |
| 29 | 20.16 | 0.02 | | | 82.70 | 39.16 | 9.26 | 26.24 | 12.69 | 48.65 | 32.91 |
| 30 | 28.06 | 0.02 | | | 93.34 | 9.26 | 15.62 | 26.26 | 93.34 | 93.34 | 37.50 |
| 31 | 44.08 | 0.02 | | | 41.02 | 52.75 | 18.26 | 24.47 | 54.80 | 44.08 | 37.67 |
| 32 | 70.32 | 0.02 | | | 12.69 | 15.23 | 9.26 | 25.13 | 3.11 | 82.70 | 32.86 |
| 33 | 15.23 | 0.02 | | | 70.32 | 24.47 | 54.80 | 85.76 | 25.13 | 20.59 | 38.77 |
| 34 | 20.59 | 0.02 | | | 28.06 | 86.37 | 13.78 | 28.38 | 44.67 | 19.50 | 37.45 |
| 35 | 28.08 | 0.02 | | | 13.78 | 17.39 | 54.80 | 42.97 | 17.00 | 20.59 | 33.51 |
| 36 | 44.67 | 0.02 | | | 54.80 | 44.67 | 18.26 | 8.88 | 22.22 | 26.24 | 39.16 |
| 37 | 82.70 | 0.02 | | | 28.08 | 18.43 | 48.65 | 44.67 | 85.76 | 28.38 | 37.55 |
| 38 | 15.62 | 0.02 | | | 12.69 | 46.69 | 12.69 | 23.04 | 38.64 | 18.43 | 33.71 |
| 39 | 22.22 | 0.02 | | | 54.80 | 45.40 | 23.04 | 18.43 | 28.06 | 9.26 | 30.75 |
| 40 | 28.38 | 0.02 | | | 24.58 | 28.06 | 20.59 | 17.00 | 27.65 | 44.08 | 32.21 |
| 41 | 45.40 | 0.02 | | | 10.81 | 18.43 | 28.08 | 10.81 | 85.76 | 52.75 | 39.03 |
| 42 | 85.76 | 0.02 | | | 19.50 | 19.54 | 24.58 | 15.23 | 23.04 | 26.24 | 37.24 |
| 43 | 17.00 | 0.02 | | | 52.75 | 15.23 | 15.62 | 32.03 | 61.22 | 86.37 | 31.86 |
| 44 | 23.04 | 0.02 | | | 59.07 | 61.22 | 17.00 | 18.26 | 42.97 | 20.16 | 36.86 |
| 45 | 32.03 | 0.02 | | | 39.16 | 19.50 | 42.97 | 28.06 | 24.58 | 41.02 | 39.26 |
| 46 | 46.69 | 0.02 | | | 52.75 | 28.38 | 3.11 | 22.22 | 18.26 | 52.75 | 37.72 |
| 47 | 86.37 | 0.02 | | | 50.39 | 24.58 | 48.65 | 15.23 | 32.03 | 15.62 | 35.80 |
| 48 | 17.39 | 0.02 | | | 48.65 | 36.37 | 9.26 | 8.88 | 70.32 | 59.07 | 35.95 |
| 49 | 24.47 | 0.02 | | | 8.88 | 28.38 | 25.13 | 18.43 | 3.11 | 8.88 | 35.74 |
| 50 | 34.98 | 0.02 | | | 24.58 | 22.22 | 44.67 | 42.97 | 59.07 | 59.07 | 35.69 |
| 51 | 48.65 | 0.02 | | | 39.16 | 45.40 | 42.97 | 85.76 | 61.22 | 44.08 | 37.57 |
| 52 | 93.34 | 0.02 | | | 20.16 | 20.16 | 82.70 | 28.08 | 41.02 | 9.26 | 36.12 |

Figure 18.5: Data and Bootstrap Worksheet

**Example 18.2.** (Exercise 18.3, page 18.15 and Exercise 18.10, page 18.23 in Companion Chapter 18.) **Spending by Shoppers.** The data on page 18.15 in the Companion Chapter shows the dollar amounts spent by 50 consecutive shoppers at a supermarket.

(a) Make a histogram of the data and observe that it is slightly skewed.

(b) Take 1000 bootstrap samples and construct a histogram of the sample means for each bootstrap sample.

(c) Make a histogram of the 1000 resamples. This is the bootstrap distribution.

(d) Calculate the bootstrap standard error.

(e) Construct a 95% bootstrap $t$ confidence interval for the population mean $\mu$ and compare with the standard one-sample $t$ confidence interval.

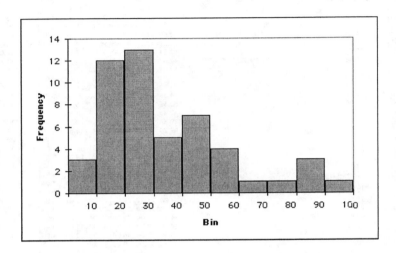

Figure 18.6: Histogram of Data

**Solution.**

Step 1. Enter the data in a column of a worksheet, for instance in cells A3:52 and then enter the equal sampling probabilities, 0.02, in the column adjacent to the right. Fig 18.5 shows a portion of our worksheet.

Step 2. As in Example 18.1 use the **Histogram Tool** to construct a histogram of the 50 data (Figure 18.6). The histogram shows skewness to the right.

Step 3. Exactly as in Example 18.1 use the **Random Number Generation Tool** to take 1000 bootstrap samples each of size 50. In our worksheet these samples are in rows 1–1000 starting in column E. We have shown a portion of the samples in Fig 18.5 and have also removed intermediate columns of data. Column BC shows the sample means of the bootstrap samples from which a histogram of the 1000 bootstrap samples was constucted using the **Histogram Tool** (Figure 18.7). We observe that the histogram looks normal.

Step 4. Since the distribution of the sample mean $\bar{X}$ from a sample of size 50 appears approximately normal we feel justified in using a confidence interval

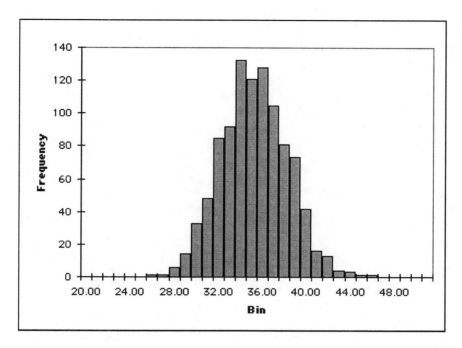

Figure 18.7: Histogram of 1000 Bootstrap Samples of $\bar{x}$

based on the $t$ (although, in view of the large sample size, this will not be distinguishable from an interval based on the $z$). The sample mean of the data is

$$\bar{x} = \texttt{=AVERAGE(A3:A52)} = 34.70$$

and from column BC we find that

$$\text{SE}_{\text{boot}} = \texttt{=STDEV(BC1:BC1000)} = 3.001$$

Therefore a 95% confidence interval for $\mu$ based on the bootstrap is

$$\bar{x} \pm t^{*}\text{SE}_{\text{boot}} = 34.70 \pm (2.009)(3.001) = (28.67, 40.73)$$

The one-sample $t$ confidence interval for $\mu$ is

$$\bar{x} \pm t^{*}\frac{s}{\sqrt{50}} = 34.70 \pm (2.009)(\frac{21.699}{\sqrt{50}} = (28.53, 40.87)$$

Note that we have obtained $s$ from the Excel formula $\texttt{=STDEV(A3:A52)}$ applied to the original data set.

The two confidence intervals are virtually identical.

# Index